Gas Metal Arc Welding

This book addresses the fundamental technologies and advancements in gas metal arc welding (GMAW) in metal joining and additive manufacturing. It delivers in-depth knowledge on various aspects of GMAW processes, including principles and mechanisms, metal transfer modes, process variations, monitoring and control, automation and flexibility of power sources, joining of similar and dissimilar materials, additive manufacturing, and metallurgical aspects. The book concludes with challenges and opportunities that would lead toward future directions in the GMAW process.

- Provides an overview of the additive manufacturing process and its capability to produce additively manufactured metallic components.
- Describes the power source and its connectivity with robots to produce complex structures.
- Discusses fundamental principles of the gas metal arc welding process.
- Explores experimental and theoretical aspects of the processes.
- Includes the most recent applications and real-time case studies along with relevant illustrations.

This book is aimed at graduate students and researchers in metallurgical engineering, additive manufacturing, and welding technology.

Gas Metal Arc Welding

Metal Joining and Additive Manufacturing

Syed Quadir Moinuddin, Murali Mohan Cheepu
and Venkata Charan Kantumuchu

CRC Press
Taylor & Francis Group
Boca Raton London New York

CRC Press is an imprint of the
Taylor & Francis Group, an **informa** business

Designed cover image: © Syed Quadir Moinuddin, Murali Mohan Cheepu and Venkata Charan Kantumuchu

First edition published 2026

by CRC Press
2385 NW Executive Center Drive, Suite 320, Boca Raton FL 33431

and by CRC Press
4 Park Square, Milton Park, Abingdon, Oxon, OX14 4RN

CRC Press is an imprint of Taylor & Francis Group, LLC

ISBN: 9781032704012 (hbk)
ISBN: 9781041314004 (pbk)
ISBN: 9781003778844 (ebk)

DOI: 10.1201/9781003778844

Typeset in Times
by Deanta Global Publishing Services, Chennai, India

Contents

Authors' Biography

Syed Quadir Moinuddin is currently Assistant Professor in the Department of Mechanical Engineering, College of Engineering at King Faisal University, Al-Hofuf, Kingdom of Saudi Arabia. Before joining, he has gained five years of experience as Assistant Professor in the Department of Mechatronics and Mechanical Engineering at Institute of Chartered Financial Analysts of India (ICFAI) Foundation for Higher Education Hyderabad. He has completed his PhD in the Department of Mechanical and Aerospace Engineering from the Indian Institute of Technology Hyderabad (IITH) and his Master's and Bachelor's degrees in the Department of Mechanical Engineering from Amrita Vishwa Vidyapeetham, Coimbatore, and Anna University, Chennai, respectively. His current areas of research are Advanced Metal Joining Processes, Metal Additive Manufacturing, Welding Metallurgy, Defect Diagnostics, and Industry 4.0. Prior to joining IITH, he was working as senior research fellow (SRF) and junior research fellow (JRF) at Indian Institute of Technology Kharagpur (IITKGP) and Welding Research Institute (WRI), Bharat Heavy Electricals Limited (BHEL) Trichy, respectively. He has also worked as Research Intern at Joining and Welding Research Institute (JWRI); Osaka University, Japan, under the Sakura Science Exchange Program. He has gained an experience of 15 years in research and 7 years of teaching experience in the field of manufacturing science. He has been awarded "Weldwell Specialty Award for Best PhD Thesis" for the year 2019 by the Indian Institute of Welding (IIW), a member society of International Institute of Welding (IIW) . He has been awarded "Excellence in Research" for the year 2015 by IITH. He has been awarded the Ministry of Human Resource Development (MHRD) fellowship by the Government of India for pursuing higher studies from 2012 to 2017. Through his research work, he has contributed to society in the form of national and international publications, namely book chapters, journal articles, and conferences. He has been an active lifetime associate member of IIW.

Murali Mohan Cheepu, born in 1988, received his Bachelor's degree in Mechanical and Production Engineering from Andhra University (GITAM College) in Visakhapatnam, India. He earned his Master's degree in Welding Engineering from the National Institute of Technology Tiruchirappalli (NITT) in India. In 2018, he was awarded his Doctoral degree from Kyungsung University in the Republic of Korea. He also carried out his Post Doctoral research at Pukyong National University, Korea, in 2019. In 2011, Dr Murali was appointed as Junior Manager at Jindal South West (JSW) Steel Ltd., where he oversaw the fabrication and maintenance of structural welding, overlay welding, and WAAM (Wire Arc Additive Manufacturing) components. Dr Murali is currently serving as the Research and Technical Sales Manager of the Welding Automation Development Center at Koweld Co., Ltd., the Republic of Korea. Prior to this, he worked as a Deputy General Manager of the Research and Technical Division at Jin Heung Tech Co., Ltd., the Republic of Korea, and Research and Project Manager at Vitzronextech Co., Ltd., the Republic of Korea, and Research Manager at Starwelds Inc., the Republic of Korea. He has several years of industry experience, most notably at JSW Steel Ltd. and Starwelds, Inc. as a Research Manager specializing in the development of WAAM processes for large-scale components, welding process design, automation, digitalization, and artificial intelligence. Dr Murali has extensive research experience in additive manufacturing, welding and joining technologies, development of welding filler materials, mechanical behavior materials, and machine learning and AI of the WAAM process. Most recently, he started working for the development of welding methods for the construction of "Artificial Sun: Future Energy: Fusion & Plasma." He has published research articles, edited books, and patents on Advances in Additive Manufacturing Processes and Automation in Welding Technologies, Green Technology, and Smart Materials for Engineering Applications, and continues to serve as the editorial board member of several journals. In 2023, Dr Murali was honored with several prestigious awards, including the Outstanding Young Manufacturing Engineer from the Society of Manufacturing Engineers (SME), USA, and the Young Fellow Award from the Korean Welding and Joining Society (KWJS), the Republic of Korea. In addition, he received the Best Researcher Award from the KWJS in 2020. Additionally, Dr Murali received the Outstanding Contributor Award from the Korean Society for Precision Engineering (KSPE), Korea, and the Outstanding Citations Award from KWJS. This recognition was further elevated in 2024 with the "A. F. Davis Silver Medal Award" from the American Welding Society, USA, and in 2025 with the "International Association of Advanced Materials (IAAM) Scientist Award" from the International Association of Advanced Materials, Sweden, highlighting Dr Murali's significant contributions and achievements in the field. Dr Murali is an IEEE senior member and a member of Korean Welding and Joining Society (KWJS), the Hong Kong Society of Mechanical Engineers (HKSME), American Welding Society (AWS), the Korean Society of Mechanical Engineers (KSME), and the Society of Manufacturing Engineers (SME).

Venkata Charan Kantumuchu currently works as the Global Quality Director at Electrex Inc., in Kansas, United States. He has more than 13 years of experience in the field of manufacturing quality, and he is a fabrication, welding, and additive manufacturing enthusiast. He has co-authored several technical articles, book chapters, and patents. He has been awarded the "Outstanding Young Manufacturing Engineer" award by the society of manufacturing engineers for the year 2022. He has published seven patents and has been granted one, in the fields of materials, energy, and technology. He has written several book chapters on additive manufacturing and welding. He was a keynote speaker at international conferences and reviewed a book on wire arc additive manufacturing fundamentals. He graduated with a Bachelor's degree in Industrial Engineering in India and has a Master's degree in Manufacturing Engineering from Bradley University, United States. He also has an MBA in Organizational Leadership from Oklahoma Baptist University. He is a senior member of IEEE and the American Society for Quality (ASQ). He is a certified quality auditor, engineer, and manager, with a Lean Six Sigma Master Black Belt.

Preface

The rapid advancement of manufacturing technologies has placed welding and additive manufacturing at the forefront of modern industrial production. These technologies enable the cost-effective fabrication of complex components while maintaining high levels of structural integrity, dimensional accuracy, and performance. Among fusion welding processes, gas metal arc welding (GMAW) has gained widespread acceptance owing to its operational flexibility, high material deposition capability, ease of integration with automation, and applicability to a broad range of metallic systems. In recent years, the scope of GMAW has expanded significantly, positioning it as a key enabling technology for wire-based additive manufacturing, particularly wire arc additive manufacturing (WAAM).

A major factor contributing to the suitability of GMAW for WAAM lies in its operator-friendly and inherently controllable process characteristics. The simultaneous use of the electric arc as the heat source and a continuously fed consumable wire as the feedstock allows for stable deposition and consistent layer formation. Process variables such as current, voltage, wire feed speed, and metal transfer behavior can be precisely regulated, enabling effective control over heat input and material transfer. As a result, GMAW-based WAAM systems demonstrate reliable arc stability, uniform bead geometry, improved surface quality, and sound metallurgical integrity. Furthermore, the extensive availability of filler wires, compatibility with robotic and automated platforms, and the adaptability of modern digitally controlled power sources make GMAW a practical, scalable, and industrially viable solution for producing high-quality additively manufactured components.

The primary aim of this book, *Gas Metal Arc Welding: Metal Joining and Additive Manufacturing*, is to provide a detailed and coherent treatment of GMAW, encompassing both its traditional applications in metal joining and its emerging role in additive manufacturing. The book integrates fundamental principles with recent technological developments, offering insights into process physics, equipment advancements, monitoring strategies, metallurgical behavior, and application-driven considerations. By doing so, it seeks to bridge the gap between academic research and industrial implementation.

Chapter 1 lays the foundation by introducing welding fundamentals and presenting a comprehensive overview of the GMAW process. It covers welding classifications, operating principles, power source characteristics, polarity selection, process parameters, filler material classification, heat generation, metal transfer mechanisms, welding symbols, and an assessment of the advantages and limitations associated with GMAW.

Chapter 2 examines metal transfer behavior in GMAW, with emphasis on both conventional and advanced transfer modes. Short-circuit, globular, and spray transfer modes are discussed alongside modern developments such as surface tension transfer (STT), cold metal transfer (CMT), regulated metal deposition (RMD), pulsed spray, rotational spray, and axial spray transfer, highlighting their governing mechanisms and practical significance.

Chapter 3 explores advanced GMAW process configurations and system-level innovations. This chapter covers twin and tandem GMAW, multi-wire arrangements including triple-wire, four-wire, and six-wire systems, cable wire GMAW, hybrid GMAW processes, and underwater welding, with discussion centered on process capabilities, performance benefits, and application domains.

Chapter 4 focuses on process monitoring and sensing techniques essential for achieving consistent weld quality and intelligent process control. Parameters such as current, voltage, wire feed rate, temperature, weld pool dynamics, acoustic emissions, and bead geometry are discussed in relation to real-time monitoring, diagnostics, and process optimization.

Chapter 5 addresses the evolution of power sources and their critical role in automation. Topics include modern inverter-based power supplies, process flexibility, integration with automated welding systems, and intelligent welding platforms capable of adaptive and data-driven control.

Chapter 6 is dedicated to the application of GMAW in metal joining. It discusses joint configurations and welding positions, joining of similar and dissimilar materials, associated metallurgical transformations, resulting mechanical properties, welding of novel materials, and modeling approaches used to predict weld performance.

Chapter 7 presents a comprehensive discussion on the use of GMAW in additive manufacturing. After introducing additive manufacturing concepts and classifications, the chapter examines WAAM using MIG/MAG, CMT, multi-wire, and hybrid GMAW approaches. Modeling strategies, microstructural evolution, mechanical behavior, and process limitations are analyzed with a focus on quality and process reliability.

Chapter 8 concludes the book by addressing current challenges and future prospects in GMAW technology. Topics include process miniaturization, welding in extreme environments such as space, advanced quality assurance methodologies, the application of machine learning, and emerging research directions that are expected to influence the next generation of GMAW-based systems.

This book is intended to serve as a reliable reference for students, researchers, academicians, and industry professionals working in welding engineering, materials science, manufacturing, and additive manufacturing. It is the author's sincere expectation that this work will contribute to knowledge dissemination, encourage technological innovation, and support the continued advancement of GMAW in both research and industrial practice.

Acknowledgment

This work was supported by the Deanship of Scientific Research, Vice Presidency for Graduate Studies and Scientific Research, King Faisal University, Saudi Arabia [KFU261133].

Foreword

Gas metal arc welding (GMAW) remains one of the most widely adopted and continuously evolving welding processes in modern manufacturing. Its versatility, high productivity, and seamless integration with automation and digital control systems have made it indispensable across industries such as automotive, aerospace, energy, shipbuilding, and heavy engineering. At the same time, ongoing scientific advances have expanded the role of GMAW beyond conventional joining, positioning it as a key enabling technology for metal additive manufacturing.

This book, *Gas Metal Arc Welding: Metal Joining and Additive Manufacturing*, presents a coherent and well-structured treatment of GMAW, combining fundamental principles with contemporary technological developments. Beginning with the basics of welding science and GMAW process characteristics, the authors systematically guide the reader through metal transfer mechanisms, advanced power sources, process monitoring techniques, and automated welding systems. The coverage reflects both academic rigor and practical relevance, making complex concepts accessible without sacrificing technical depth.

Particularly noteworthy is the discussion on advanced metal transfer modes, multi-wire and hybrid GMAW configurations, and intelligent process monitoring areas that are critical for achieving high-quality, repeatable welds in industrial environments. The book further extends its scope to wire arc additive manufacturing (WAAM), addressing process variants, metallurgical evolution, mechanical performance, and modeling approaches, thereby aligning GMAW with the future of digital and additive manufacturing.

The final chapters offer a forward-looking view by discussing current challenges and new opportunities, including miniaturized welding systems, space applications, quality assurance, and the integration of machine learning and artificial intelligence. These insights emphasize the evolving nature of GMAW research and its increasing role in next-generation manufacturing systems.

Overall, this book serves as a valuable reference for graduate students, researchers, practicing engineers, and industry professionals. By effectively bridging academic research and industrial practice, it makes a meaningful contribution to the advancement of welding and additive manufacturing technologies.

Professor Dr Mohammed Saber
Department of Mechanical Engineering
College of Engineering
King Faisal University
Al-Ahsa – 31982
Saudi Arabia

1 Introduction to Gas Metal Arc Welding

1.1 WHAT IS WELDING?

Welding is a fundamental fabrication process used to join two or more materials—typically metals—by creating a strong, permanent bond through the application of heat, pressure, or both. The materials may be similar or dissimilar, and the process may or may not involve a filler material that assists in forming the joint. Welding is widely regarded as a cost-effective and efficient technique for assembling components in a vast range of industries, including aerospace, automotive, shipbuilding, construction, and defense [1, 2]. According to the American Welding Society (AWS), welding is defined as "a process that produces coalescence of materials by heating them to the welding temperature, with or without the application of pressure, and with or without the use of filler metal" [3, 4]. Similarly, The Welding Institute (TWI-UK) describes welding as "a fabrication method that fuses two or more parts using heat, pressure, or both, to form a joint as the parts cool." This joint, formed through coalescence, must possess sufficient strength to perform under service conditions. The effectiveness of a welded joint is influenced by several factors, including the type of materials being joined, the welding technique employed, the thermal cycle involved, and the interaction of heat and mechanical forces at the interface. Welding is not merely about melting and joining—it is about achieving a durable, reliable bond that can withstand structural and operational demands. Given its importance, understanding welding requires a comprehensive examination of its principles, classifications, heat sources, modes of metal transfer, and the advantages and limitations of different welding methods. This chapter introduces the fundamentals of welding with a particular focus on the gas metal arc welding (GMAW) process, detailing its working principles, key parameters, benefits, limitations, and recent advancements.

1.2 CLASSIFICATIONS OF WELDING

Welding processes can be broadly classified into two main categories: fusion welding and non-fusion welding as shown in Figure 1.1. This classification is based on the fundamental mechanism through which the materials are joined—whether by melting or without melting.

In fusion welding, the base materials at the joint interface are heated to a temperature above their melting point, resulting in the formation of a molten pool. This molten material then solidifies upon cooling to form a strong metallurgical bond

DOI: 10.1201/9781003778844-1

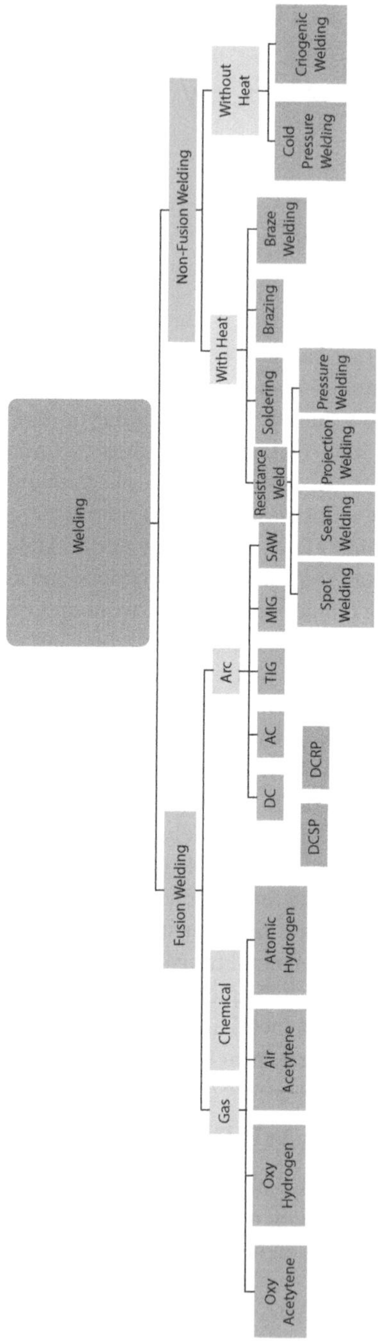

FIGURE 1.1 Classification of welding process [5].

between the components. In many fusion welding processes, a filler metal is added to the weld pool to facilitate joint formation and enhance mechanical strength. These processes do not require the application of external pressure during welding. Fusion welding is commonly used for a wide range of materials, including ferrous and non-ferrous metals, and is favored in applications that require strong, continuous joints. Examples of fusion welding processes include gas metal arc welding (GMAW), shielded metal arc welding (SMAW), gas tungsten arc welding (GTAW), submerged arc welding (SAW), plasma arc welding (PAW), and laser beam welding (LBW). The source of heat in fusion welding may be an electric arc, plasma, laser, or combustion flame.

On the other hand, non-fusion welding, or solid-state welding, achieves coalescence without melting the base materials. Instead, the materials are joined by applying pressure, and in many cases, heat is used to soften but not melt the interface. This category of welding relies on atomic bonding, plastic deformation, and diffusion at the interface to produce a solid joint. Since the materials are not melted, the original microstructure is largely retained, minimizing thermal distortion, residual stresses, and potential metallurgical defects. These advantages make solid-state welding particularly suitable for joining dissimilar metals, heat-sensitive alloys, and precision components. Common examples of non-fusion welding processes include friction stir welding (FSW), ultrasonic welding (UW), explosion welding, forge welding, diffusion welding, and cold welding.

Understanding this fundamental classification is essential for selecting appropriate welding processes for different applications, especially when considering factors such as joint strength, material compatibility, heat input, and mechanical performance. In the following sections, special emphasis will be placed on fusion welding, particularly the GMAW process, due to its widespread industrial relevance and adaptability.

1.3 PRINCIPLES OF GMAW

Gas metal arc welding (GMAW) is a fusion welding process in which an electric arc is established between a continuously fed consumable wire electrode and the workpiece. The arc generates intense heat, which melts both the electrode and the base metal, creating a molten weld pool that solidifies to form a strong, permanent joint. This process is widely used in manufacturing and fabrication industries due to its high deposition rate, automation compatibility, and suitability for a variety of materials and thicknesses.

The fundamental principle of GMAW lies in the simultaneous melting of the electrode and the base metal by the heat of the arc. The electrode wire, typically made of the same or similar material as the base metal, is continuously fed from a spool through a wire-feeding system and welding torch. As the wire reaches the arc zone, it melts and transfers into the weld pool. The welding torch also supplies a shielding gas—usually an inert gas like argon or a mixture of argon and carbon dioxide—which protects the molten weld pool and electrode from atmospheric contamination such as oxygen, nitrogen, and moisture. Without this shielding, the weld quality would degrade due to porosity, oxidation, or slag inclusions.

A simplified schematic of the GMAW setup is shown in Figure 1.2. The main components of the system include:

- A power source (usually constant voltage DC);
- A wire feed mechanism to supply the electrode;
- A welding torch that delivers both the electrode and shielding gas;
- A shielding gas cylinder and regulator system;
- Grounding and control units to ensure stable arc performance.

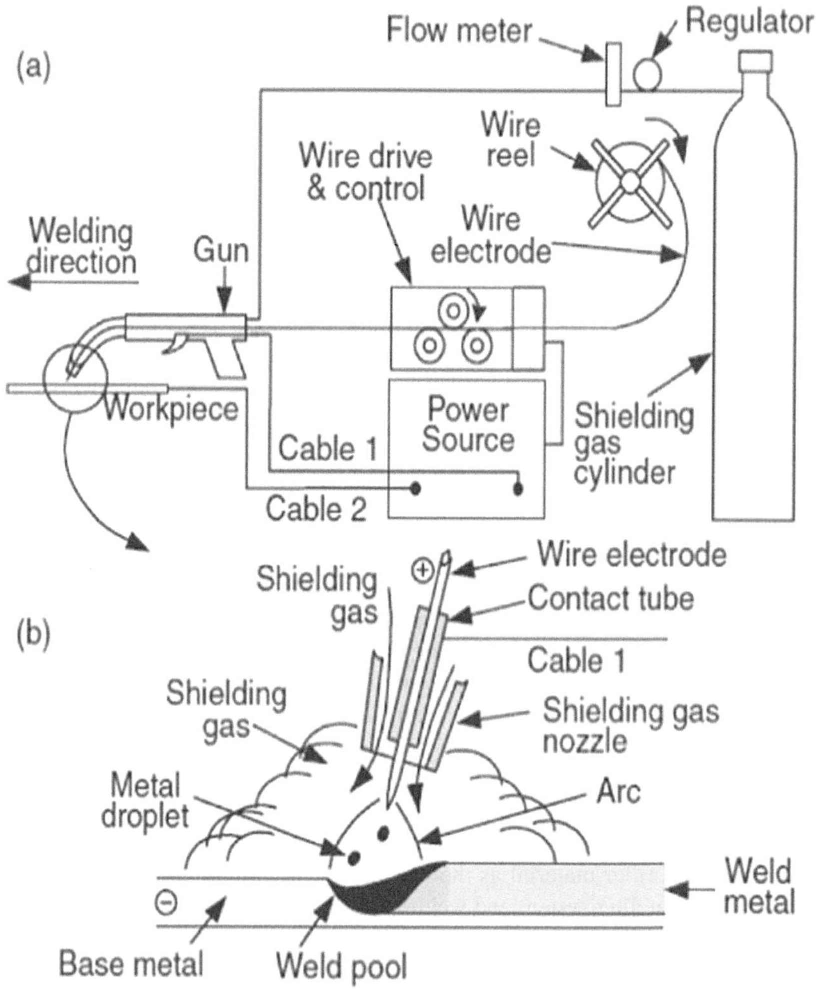

FIGURE 1.2 GMAW process: (a) overall and (b) enlarged welding area [6].

The arc is initiated when the electrode wire comes into contact with the workpiece and a short circuit is created. Once the arc is struck, the shielding gas becomes ionized, forming a conductive plasma channel. This ionized path allows the continuous flow of current, maintaining the arc and sustaining the weld pool as the torch moves along the joint line. The movement of the torch, in coordination with the feed rate of the wire and other parameters, ensures the progressive formation of the weld bead.

A critical factor in the effectiveness of the GMAW process is arc efficiency, which refers to how much of the supplied electrical energy is effectively utilized in melting the base and filler metal. A portion of the input energy is inevitably lost through convection, radiation, and heat absorption by the torch, electrode, and surrounding atmosphere. The thermal efficiency or arc efficiency (η_a) is quantitatively defined as

$$\eta_a = P_n / P_g \qquad \text{.......} (1.1)$$

Here η_a = arc efficiency or thermal efficiency
P_n = net power input transferred to the plate
P_g = gross power input to arc from the source

Typical values of arc efficiency vary depending on the welding process. For GMAW, efficiency generally ranges from 0.60 to 0.85, which is higher than that of processes like GTAW, which may range from 0.22 to 0.80 [1]. Efficient energy transfer is crucial for consistent penetration depth, bead geometry, and mechanical properties of the weld. Figure 1.3 illustrates the energy distribution during the GMAW process, highlighting both useful and lost components of energy.

In addition to energy management, torch positioning, arc length, travel speed, and wire feed rate are key parameters that influence the quality and consistency of the

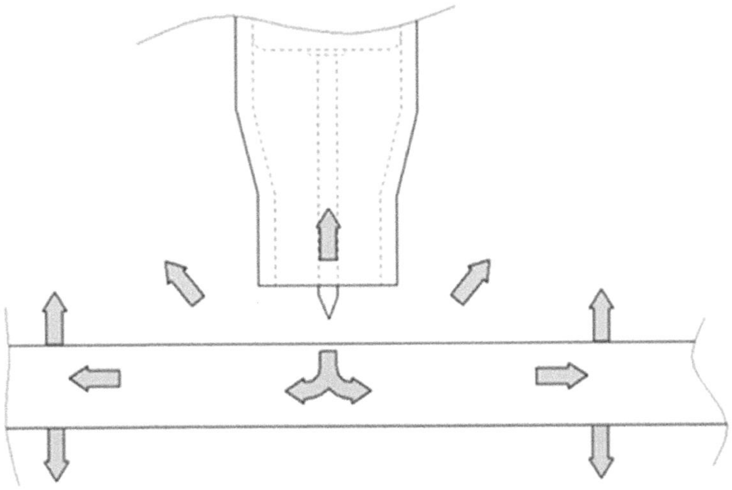

FIGURE 1.3 Heat transfer around a torch [7].

weld. Deviations in these parameters can lead to defects such as undercutting, lack of fusion, or excessive spatter. Therefore, a thorough understanding of process physics and control strategies is essential for achieving reliable welds in both manual and automated GMAW operations.

1.3.1 Power Characteristics

Welding processes including GMAW require reliable and tailored power sources capable of generating sufficient energy to produce high-quality welds. In arc welding, the power source plays a pivotal role in establishing and maintaining a stable arc. It delivers energy in the form of low-voltage and high-current output, which is essential for the formation and maintenance of the welding arc [8]. GMAW power sources are typically classified as constant voltage (CV) or constant current (CC) systems, operating on either alternating current (AC) or direct current (DC). For GMAW, DC with constant voltage characteristics is predominantly used, offering smoother arc behavior and better metal transfer control. The usual current rating required for industrial welding operations ranges from 70 to 400 amperes, with duty cycles of 60% to 80% depending on the application and machine capacity.

1.3.1.1 Voltage Ranges and Arc Behavior

Industrial power systems typically supply 240 V or 480 V, which is transformed by the welding power source. The open-circuit voltage (OCV) required to initiate the arc generally falls within the range of 50–80 V. Once the arc is struck, the working or arc voltage stabilizes in the range of 10–30 V, depending on the electrode extension, arc length, shielding gas, and base material. The arc itself acts as a dynamic impedance in the circuit. This impedance is a function of temperature, particle density, and mobility of charged species within the arc plasma. As temperature rises, ionization increases, which reduces arc impedance. The spatial distribution of these charged particles both radially and axially significantly influences arc stability and power transfer efficiency.

1.3.1.2 Voltage Zones in Welding Arc

The electric arc in GMAW consists of three distinct regions that contribute to the overall arc voltage (V)

- Cathode drop region (V_c): Zone near the cathode where electrons are emitted.
- Plasma region (V_p): The main arc column where ionized gas conducts current.
- Anode drop region (V_a): Near the workpiece where electrons are absorbed.

These regions collectively account for the total arc voltage. The arc power (P), or the heat energy generated per unit time, is calculated using

$$\text{Power of the arc } (P) = (V_c + V_p + V_{a)} * I \ldots\ldots \quad (1.2)$$

Where

P = arc power (W)
I = arc current (A)
V_c, V_p, V_a = Voltage drops across respective arc regions

Figure 1.4 shows a schematic of these voltage zones, while the energy flow and the movement of charged particles take place within the arc plasma.

1.3.1.3 Functional Requirements of GMAW Power Sources

To ensure efficient and safe welding operations, the power source in the gas metal arc welding (GMAW) system must fulfill several critical functional requirements. First and foremost, the welding power source must electrically isolate the welding circuit from the main power supply. This isolation is essential not only for the safety of the operator but also for protecting the welding equipment from voltage fluctuations and electrical faults in the main line.

Furthermore, the power source must be capable of delivering precise and consistent output in terms of voltage and current across a wide range of welding applications. This regulation capability ensures that variations in arc length or electrode extension do not disrupt the welding process. In particular, the output voltage–current (V–I) characteristics of the power source must match the demands of the arc system. For GMAW, this typically means a constant voltage (CV) characteristic, which facilitates self-regulation of the arc length during metal transfer.

Another critical function of the power source is its ability to handle low-impedance loads, such as the short-circuiting conditions encountered in certain metal transfer modes like short-circuit GMAW. Additionally, modern power sources should provide options for both AC and DC outputs, although DC is predominantly used in GMAW for its arc stability and cleaner welds. Advanced power sources may

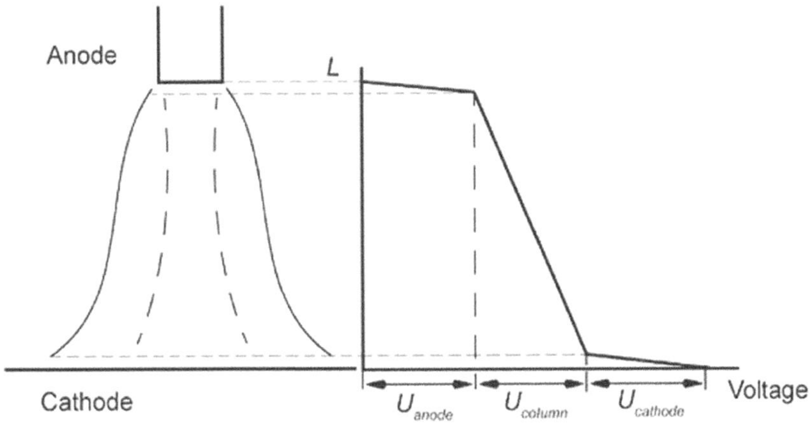

FIGURE 1.4 Representation of three voltage drop zones in the electric arc [9].

also include digital interfaces, parameter monitoring, and adaptability for robotic or automated welding systems.

1.3.1.4 Selection Criteria of Power Source

The selection of an appropriate power source for gas metal arc welding (GMAW) involves a careful evaluation of multiple practical and technical factors to match the specific requirements of the application and working environment. A critical consideration is the cost—including the initial investment, operational efficiency, and long-term maintenance. Industrial applications typically demand more robust, feature-rich units, whereas smaller workshops or field operations may prioritize cost-effective and portable systems. Another vital factor is compatibility with the available electrical infrastructure. In many workshops and construction sites, 220 V single-phase power may be the standard, necessitating portable power sources that can deliver stable output under variable input conditions. In contrast, large-scale manufacturing facilities often use 480 V three-phase power to support higher current capacities needed for continuous, high-deposition welding operations.

The type of current (AC or DC) is a major determinant in power source selection. In GMAW, DC power sources are generally preferred due to their superior arc stability, deeper and more consistent penetration, and better control over heat input. In comparison, as represented in Table 1.1, AC power sources, while more economical and requiring less maintenance, tend to produce fewer stable arcs and more spatter, making them less suitable for GMAW.

In addition to current type, output characteristics such as constant voltage (CV) or constant current (CC) must align with the GMAW process. GMAW relies predominantly on constant voltage power sources to maintain a steady arc length and ensure consistent metal transfer. The *V–I* characteristic of the power source must be flat enough to accommodate fluctuations in arc length while maintaining stable arc behavior. Other selection considerations include the current rating, duty cycle, and remote-control compatibility for automated or robotic welding systems. A power source with remote adjustability is especially valuable in industrial environments where frequent parameter tuning is necessary during complex or long welding sequences. Furthermore, the ability of the power source to deliver a stable arc under

TABLE 1.1
Comparison of Power Sources

Feature	Alternating Current (AC)	Direct Current (DC)
Arc stability	Less stable	More stable
Equipment cost	Lower	Slightly higher
Maintenance	Lower	Moderate
Weld quality	May cause arc interruptions	Smoother, cleaner welds
Polarity control	Not possible	Easily achievable (DCEP, DCEN)

varying load conditions, particularly during short-circuiting transfer, is essential to achieve sound welds without defects.

In summary, selecting the right power source for GMAW requires balancing cost, power availability, portability, control features, and process compatibility. Given these factors, DC power sources with constant voltage output remain the standard in most GMAW applications, offering optimal performance, flexibility, and weld quality.

1.3.2 Polarity

Polarity plays a crucial role in arc welding processes, particularly in gas metal arc welding (GMAW), where it directly affects arc stability, heat distribution, penetration depth, and overall weld quality. Polarity refers to the direction of electron flow in the welding circuit and is determined by how the electrode (wire) and workpiece are connected to the power source terminals. GMAW typically uses direct current (DC), and depending on the application, the polarity can be either direct current electrode positive (DCEP) or direct current electrode negative (DCEN).

1.3.2.1 Direct Current Electrode Negative (DCEN)

Also known as straight polarity, DCEN involves connecting the electrode to the negative terminal and the workpiece to the positive terminal of the power supply, as shown in Figure 1.5(a). In this case, electrons flow from the electrode to the workpiece, and much of the heat (about 70%) is focused on the workpiece surface. While this polarity theoretically seems advantageous for heating the base material, DCEN in GMAW often leads to:

- Poor arc stability
- Inadequate penetration
- Higher chances of lack of fusion
- Reduced metal transfer efficiency, especially in spray or globular transfer modes

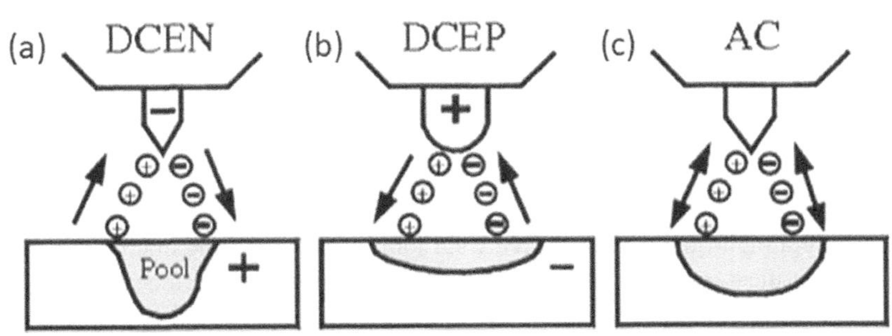

FIGURE 1.5 Schematic diagrams of (a) straight polarity and (b) reverse polarity [10].

Due to these limitations, DCEN is seldom used in standard GMAW operations. It is occasionally considered for very thin materials or specific niche applications, such as surfacing or cladding, where reduced penetration is desirable. Some research also explores its use for hard-to-weld materials like cast iron, but it generally requires significant control over process parameters.

1.3.2.2 Direct Current Electrode Positive (DCEP)

Also referred to as reverse polarity, DCEP is the standard and most widely used polarity in GMAW applications. In this configuration, the electrode is connected to the positive terminal of the power source, and the workpiece is connected to the negative terminal as shown in Figure 1.5(a). This setup causes electrons to flow from the workpiece (negative) to the electrode (positive). However, most of the arc heat (approximately 70%) is concentrated at the positive terminal, that is, on the electrode tip while the remaining 30% is delivered to the workpiece. This concentration of heat at the electrode tip facilitates:

- Stable arc initiation and maintenance
- Deeper penetration into the workpiece
- Improved weld bead fusion

Thus, DCEP is ideal for welding thicker sections of metals and is commonly used with solid and metal-cored wires in both ferrous and non-ferrous materials. It is also effective when welding materials with high thermal conductivity, such as aluminum and copper alloys. However, DCEP also presents certain challenges:

- Increased spatter formation, which can affect surface finish and post-processing requirements.
- Higher electrode melt-off rates, requiring efficient wire feeding systems.
- Potential arc blow in magnetic materials due to the electromagnetic field interaction.

Despite these drawbacks, DCEP remains the preferred polarity for most GMAW operations due to its robust arc characteristics and deeper penetration capabilities.

A clear distinction between the two polarities is essential for selecting the right configuration for a given application. DCEP is generally preferred for its versatility and reliability in most GMAW scenarios, while DCEN is limited to niche applications due to its inconsistent performance as represented in Table 1.2.

1.3.3 Process Parameters

Achieving a high-quality, defect-free weld joint in GMAW fundamentally depends on the identification and optimization of various process parameters. These parameters not only influence the weld bead geometry but also control fusion characteristics, metal transfer modes, and overall mechanical performance of the weldment.

TABLE 1.2
Differences between Straight Polarity and Reverse Polarity

Characteristic	DCEN	DCEP
Type of current	Direct current (DC)	Direct current (DC)
Polarity type	Straight polarity	Reverse polarity
Position of electrode	Negative	Positive
Flow of electrons	Electrode to work	Work to electrode
Flow of ions	Work to electrode	Electrode to work
Penetration depth range	Deep	Narrow
Suitable methods	TIG and stick welding, etc.	MIG and stick welding, etc.
Distribution of heat	One-third to electrode and two-thirds to work	Two-thirds to electrode and one-third to work

Source: Reference [6].

TABLE 1.3
List of Various Process Parameters of the GMAW Process

Primary Parameters	Secondary Parameters	Fixed Parameters
Current	Shielding gas and composition	Workpiece geometry
Voltage	Flow rate of shielding gas	Workpiece thickness
Wire feed	Electrode type	Physical properties
Torch speed	Electrode diameter	
	Torch angle/electrode orientation	
	Stick out (CTWD) value	

The welding arc is a complex, transient system where thermal, electrical, and fluid flow phenomena interact. Thus, selecting appropriate operating variables is essential for ensuring arc stability, consistent deposition, and desired weld quality. The key parameters are broadly classified into fixed, primary, and secondary categories, as listed in Table 1.3. This classification helps in determining the factors that must be set before welding and that can be adjusted in real time for quality control.

A variety of process parameters critically influence the quality, geometry, and structural integrity of welds in GMAW. Among these, arc voltage, welding current, and welding speed are primary parameters that directly determine the amount of heat input, arc stability, and the resulting bead shape and penetration. Electrode stick-out (the distance from the contact tip to the workpiece) also affects current density and heat distribution, while the diameter of the electrode impacts deposition rates and arc concentration. Wire feed rate is closely linked with current levels and deposition efficiency, playing a central role in maintaining a stable arc and consistent metal transfer. In addition, the type and flow rate of shielding gas significantly affect arc characteristics, spatter levels, and weld pool protection from atmospheric contamination. Proper adjustment of these parameters is essential to achieve a desired bead profile, control spatter, minimize weld defects, and optimize productivity. When pulsed GMAW is employed, additional variables such as peak current, background current, pulse duration, and frequency must also be fine-tuned to ensure stable metal transfer, reduce heat input, and achieve smooth, defect-free welds.

Pulsed GMAW is a variant of conventional GMAW, especially suited for out-of-position welding, thin materials, and applications requiring low heat input. This technique modulates the welding current between a high (peak) and low (background) level in a controlled waveform, allowing spray transfer at lower average current values. The important parameters in pulsed GMAW include:

- *Peak Current* (I_p): The maximum current level during the pulse cycle. It is responsible for forming and detaching the molten droplet, thus ensuring spray transfer.
- *Background Current* (I_b): A lower current level that maintains the arc without contributing to droplet detachment. It ensures arc stability during non-transfer phases.
- *Pulse Width* (T_p): Duration of the peak current in one cycle. This affects droplet formation time and heat input.
- *Background Duration* (T_b): The time during which the current remains at background level before the next pulse.
- *Pulse Frequency* (F): Number of pulses per second. Higher frequency results in finer droplet transfer but may require precise synchronization with wire feed rate.
- *Pulse Cycle Time* (T): Total duration of one full current cycle ($T_p + T_b$).
- *Load Duty Cycle* (D): The ratio of pulse width to the total cycle time. This governs the effective thermal input per unit time.

$$D = \left(\frac{T_p}{T_p + T_b} \right) \times 100\% \qquad \text{……. (1.3)}$$

Based on the peak and background current, different waveforms can be drawn out. Figure 1.6(a) shows a square waveform, which is the simplest and most widely used due to its clear and abrupt transitions between peak and base currents, providing consistent droplet detachment and stable arc characteristics [11]. Figure 1.6(b) presents a trapezoidal waveform, which introduces controlled rise and fall times to reduce abrupt current changes. This helps minimize arc instability, spatter, and electromagnetic interference [12]. Figure 1.6(c) depicts an exponential waveform, wherein current changes follow a gradual rise and decay curve, closely mimicking natural electrical and thermal transients. This waveform can be beneficial in achieving smooth arc initiation and stable transfer with reduced thermal stress.

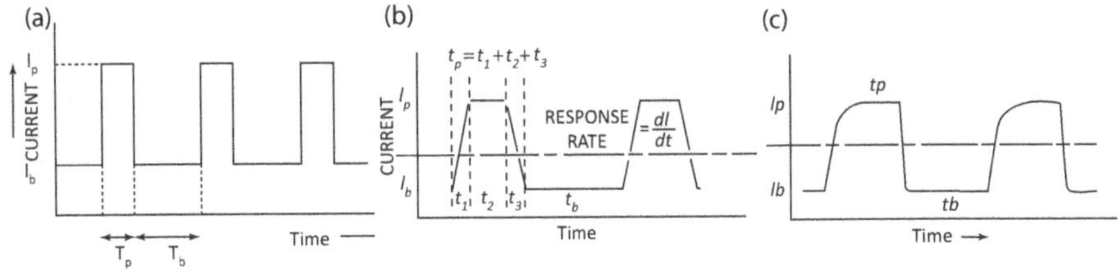

FIGURE 1.6 Signal profiles: (a) square signal [11], (b) trapezoidal signal [12], and (c) exponential signal [12].

1.3.3.1 Type of Shielding Gas and Flow Rate

Shielding gases play a vital role in protecting the molten weld pool from contamination by atmospheric gases during welding. Their function extends beyond simple protection; they significantly influence the arc characteristics, metal transfer mode, and ultimately, the microstructure and mechanical properties of the weld metal. A thorough understanding of how different shielding gases affect various base materials and welding conditions is essential for achieving optimal weld quality. Over the years, extensive research and experimental studies have provided valuable insights into the behavior and selection of shielding gases in GMAW [13]. The primary purpose of a shielding gas is to displace atmospheric air around the weld zone, thereby preventing contamination by oxygen, nitrogen, and water vapor elements that can cause oxidation, porosity, and embrittlement in the weld metal. To be effective, a shielding gas must meet several critical requirements [11]:

- It must support the generation and maintenance of a stable, concentrated arc plasma and a consistent arc root on the electrode and workpiece.
- It should promote smooth detachment and transfer of molten metal droplets from the electrode to the weld pool, minimizing spatter.
- It must protect the electrode tip, molten weld pool, and adjacent weld head components from oxidation and other adverse chemical reactions.

Carbon dioxide (CO_2) is a commonly used shielding gas in GMAW, known for its oxidizing properties under the high-temperature arc environment. The CO_2 dissociates and reacts to create a strongly oxidizing atmosphere, which can lead to the burning loss of critical alloying elements such as chromium (Cr), nickel (Ni), manganese (Mn), and silicon (Si) in the weld metal. This oxidation reduces the overall strength and toughness of the deposited weld metal [14]. Consequently, welding with pure CO_2 often results in increased oxygen content and the formation of oxide inclusions that negatively impact mechanical properties.

To mitigate these issues, ternary shielding gas mixture combinations of inert gases such as argon (Ar), with small amounts of CO_2 and nitrogen (N_2) are frequently used. These mixtures reduce the oxidizing effect by lowering the oxygen content in the welding atmosphere, which in turn reduces oxidation losses of alloying elements and spatter generation. This results in a cleaner weld metal with improved purity and mechanical properties. However, it is important to note that under the arc, nitrogen can dissociate and dissolve into the weld pool, increasing the nitrogen content of the deposited metal. This can be beneficial or detrimental depending on the alloy system and desired properties [15]. Beyond composition, the flow rate of shielding gas is also a critical parameter. It must be sufficient to displace atmospheric gases effectively without causing turbulence or excessive gas consumption. Improper flow rates can result in porosity or weld defects due to inadequate shielding or gas turbulence. Typical flow rates vary depending on the welding environment, joint design, and gas composition, and they must be optimized for each application.

To summarize, understanding the interplay between shielding gas type, composition, and flow rate is crucial to controlling weld quality and ensuring the desired metallurgical and mechanical properties. A detailed analysis of how these parameters influence the GMAW process and weld outcome is presented in Table 1.4.

1.3.4 Classification of Fillers

In the gas metal arc welding (GMAW) process, filler materials are critical in achieving high-quality, reliable weld joints. The choice of filler material not only influences the mechanical integrity but also dictates the chemical compatibility and corrosion resistance of the welded component. Selecting an appropriate filler material requires matching its chemical composition closely with that of the base metals to be joined. This ensures that the resulting weld metal meets the necessary mechanical properties and environmental resistance demanded by the application. During welding, the weld metal solidifies rapidly, typically forming a dendritic microstructure. This weld metal is a mixture of the melted base metal and the filler material, the latter being particularly essential when joining thicker sections or dissimilar metals. The filler metal contributes alloying elements that help maintain or enhance the desired properties in the weld region, compensating for any dilution effects from the base metal [16].

TABLE 1.4
Effect of Primary Parameters Used for the GMAW Process

Change in Parameter	Outcome
Increase in current	Increase in bead deposition, penetration, and fusion of substrate
Increase in welding speed/ travel speed	Decrease in bead width and penetration
Increase in voltage	Decrease in bead width and reinforcement Increase in bead width Increase in spatter
Decrease in gas flow rate	Increase in blow holes and porosity
Decrease in stick-out value	Increase in final bead quality and penetration Decrease in the weld deposit with high arc intensity

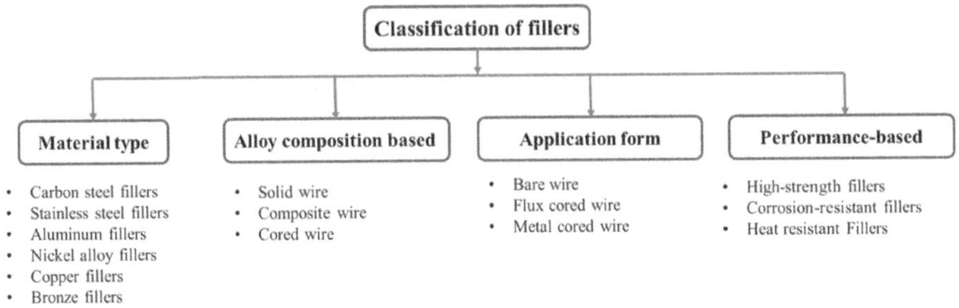

FIGURE 1.7 Classification of fillers.

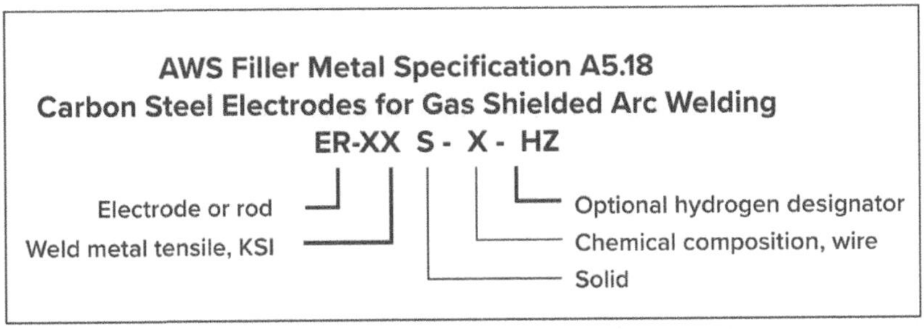

FIGURE 1.8 Filler wire specification. (Courtesy: AWS).

Filler wires used in GMAW are broadly classified based on several criteria, including the base material type (carbon steel, stainless steel, aluminum, etc.), chemical composition, and wire form (solid, flux-cored, metal-cored). This classification is visually summarized in Figure 1.7, which presents the categories and subcategories to aid in selecting the correct filler for specific applications. Furthermore, Figure 1.8 details the standardized designation system for wire electrodes commonly used as filler materials in GMAW, providing clarity on how wire specifications correspond to their composition and application suitability. Operational parameters such as current range and wire feed speed are closely related to the filler wire diameter, impacting deposition rate and weld quality. Table 1.5 offers a guideline relating current ranges with wire feed speeds for various filler wire diameters, although these ranges can vary depending on factors like shielding gas composition and base metal characteristics. Proper control and adjustment of these parameters are vital for ensuring consistent weld penetration, bead geometry, and mechanical performance.

1.3.5 Heat Generation

Heat generation and its management are fundamental in welding, as they directly influence the microstructure, mechanical properties, and dimensional stability of the welded components. The cooling rate following welding is particularly critical

TABLE 1.5
Relation of Current Range with Size and Wire Feed Rate Fillers

S. No.	Diameter (mm)	Current Range (A)	Wire Feed Rate (m/min)
1	0.6	40–100	2–5
2	0.8	40–150	3–6
3	1.0	100–280	3–12
4	1.2	120–350	4–18

Courtesy: Weldingandndt.com.

TABLE 1.6
Values of Arc Efficiency for Common Fusion Welding Methods

S. No.	Type of Fusion Welding	Arc Efficiency
1	Gas metal arc welding	0.90
2	Shielded metal arc welding	0.90
3	Submerged arc welding	0.95
4	Gas tungsten arc welding	0.70
5	Flux cores arc welding	0.90

because it dictates the phases formed within the weld and heat-affected zones (HAZ). Faster cooling typically promotes the formation of harder and more brittle phases, such as bainite and martensite, which can increase hardness but may reduce toughness. Conversely, excessive heat input increases the risk of deformation, distortion, and undesirable grain growth, adversely affecting mechanical performance. The heat input during fusion welding processes is quantitatively estimated by the following equation [16]:

$$\text{Heat input}\left(\text{J}/\text{mm}\right)=\frac{\text{V*I}}{\text{S*}\eta} \qquad \text{.......} \ (1.4)$$

V: Welding voltage (volts),
I: Welding current (amperes),
S: Welding speed (mm/s),
η: Arc efficiency (related to the heat transfer capability of the welding method).

Arc efficiency values (η) for various fusion welding methods are listed in Table 1.6 as examples.

1.3.5.1 Generation of Arc

The welding arc is generated by an electric discharge created when current flows from the cathode to the anode, producing intense heat. The arc gaps, the space between the electrode and workpiece, contains a column of charged particles (ions and electrons) that facilitates current flow due to its high electrical conductivity. Several mechanisms, such as thermal emission, field emission, and secondary emission, generate these charged particles. Electrons are emitted from the cathode and accelerated toward the anode by the potential difference, colliding with gaseous molecules in the arc gap. These collisions ionize the gas molecules, creating ions and additional electrons. Ions migrate toward the cathode while electrons move toward the anode, as illustrated in Figure 1.4(b). Since ions are much heavier, their current contribution is roughly 1% of the electron current. Heat is produced when the high-energy electrons strike the anode, converting electrical energy into thermal energy. The arc gap thus acts as a resistive load, and the heat generated is the product of arc voltage and welding current.

1.3.5.2 Arc Characteristics

The stability and behavior of the welding arc are essential factors that directly affect weld quality, bead appearance, and metallurgical characteristics. The relationship between arc voltage and current plays a crucial role in understanding how the arc behaves under varying conditions. Figure 1.9 illustrates this relationship through two different characteristic curves. Figure 1.9(a) presents the typical arc characteristics, where the voltage–current (V–I) relationship is nonlinear. Unlike a pure resistive load, the arc does not strictly follow Ohm's law. As depicted, slight variations in arc voltage can lead to a wide range of current changes, and notably, the curve does not pass through the origin. The slope of this curve depends on the welding parameters such as arc length, base material, and shielding gas composition. A steeper slope indicates a shorter arc and more concentrated heat, while a flatter slope corresponds to a longer arc and a more diffused heat zone. Figure 1.9(b) illustrates the characteristics of a CV power supply, which is commonly used in the GMAW process. In this setup, a small change in arc length results in a significant change in current but only a minor change in voltage. This inherent self-regulating behavior is beneficial for maintaining a consistent arc length. When the arc shortens, the current increases, accelerating the wire burn-off rate and lengthening the arc back to the present value. Conversely, if the arc becomes too long, the current decreases, slowing the burn-off rate and shortening the arc. This feedback mechanism maintains process stability and uniformity in metal transfer and penetration, which is particularly advantageous in automated and robotic GMAW applications.

1.3.5.3 Heat Transfer and Effects on Microstructure

During fusion welding, not all the electrical energy converts to useful heat in the weld zone; some heat is lost to the atmosphere and adjacent cold regions of the workpiece, affecting the arc efficiency. Excessive heat input can cause grain coarsening in the weld metal and HAZ, leading to reduced toughness and increased residual stresses

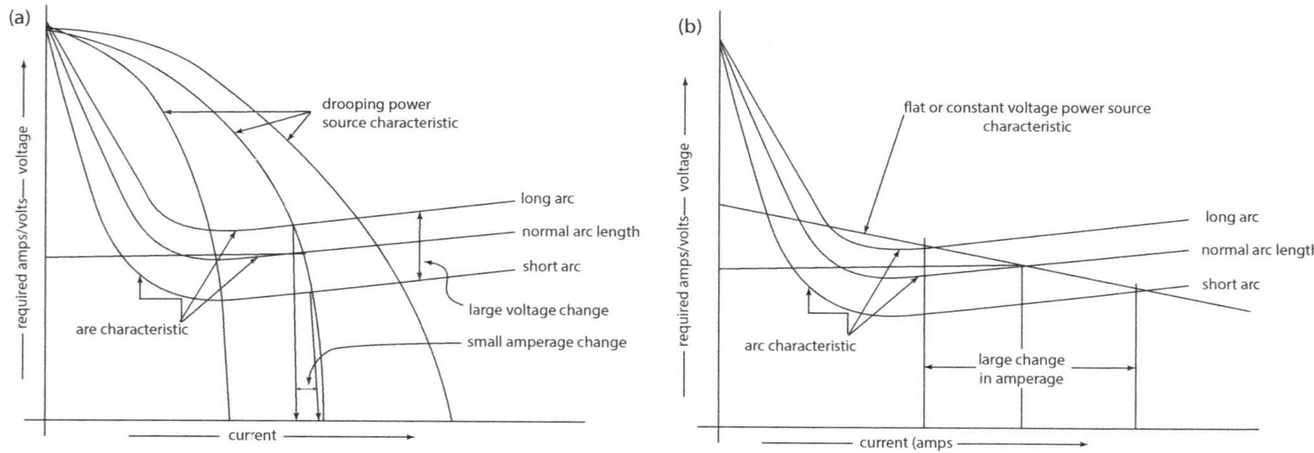

FIGURE 1.9 Arc characteristics for (a) constant current and (b) constant voltage. (Courtesy: TWI [17]).

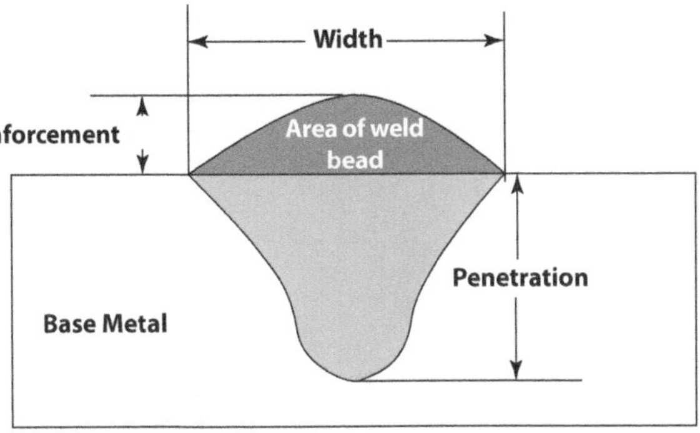

FIGURE 1.10 Main regions in fusion welding [20].

and distortion, which can shorten component life [18]. The heat source in welding directly influences the metallurgical transformations in the weld zone. Metals and alloys often undergo phase transformations when exposed to welding temperatures, altering the microstructure and mechanical properties [19]. The welded joint typically comprises three distinct regions (see Figure 1.10):

- Base Metal: Unaffected by heat, retaining its original microstructure and properties.
- HAZ: Experiences thermal cycles that may cause phase transformations, grain growth, or precipitation changes without melting.
- Weld Metal: The region of melted and solidified metal formed by the fusion of filler and base metals.

The extent of microstructural changes depends on base metal composition, filler material, welding process, and thermal cycle experienced during welding. In many ways, the weld zone can be considered akin to a casting, where rapid solidification rates influence microstructure development.

1.4 METAL TRANSFER MODES IN GMAW

In the gas metal arc welding (GMAW) process, molten metal droplets are transferred from the tip of the consumable wire electrode into the weld pool through the arc. This metal transfer occurs in several distinct modes, each influencing weld quality, heat input, bead shape, and spatter levels. The electric arc, generated between the wire electrode and the metal workpiece, facilitates this transfer. The metal transfer mode significantly affects weld bead penetration, metal deposition rate, and the thermal profile of the weld zone. These modes are governed by various process parameters, including welding current and voltage, stick-out length, electrode material, shielding gas composition, wire feed speed, and welding speed, as well as the metallurgical composition of the base material [21, 22].

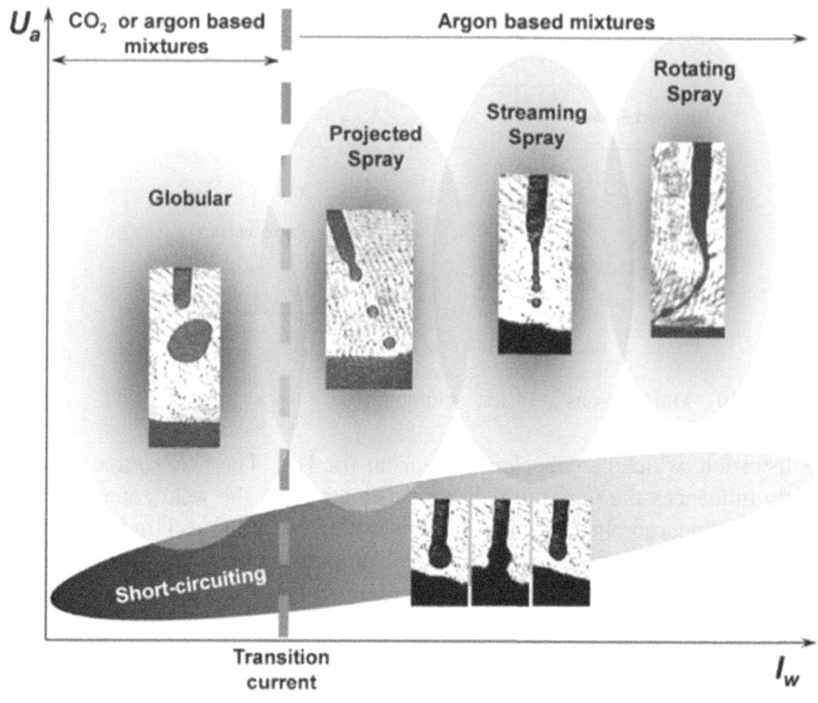

FIGURE 1.11 Classification of transfer modes [23].

Metal transfer modes describe the mechanisms by which molten droplets detach from the electrode and reach the weld pool. Accurate identification and control of these modes are essential for automated process monitoring and quality assurance. Key factors that influence these modes include droplet size, growth rate, and detachment frequency [23]. Cullen et al. [24] investigated the automatic classification of metal transfer modes in GMAW using machine learning techniques. Signals such as arc acoustics, welding current, voltage, and shielding gas flow rate were analyzed in both time and frequency domains. A support vector machine (SVM) classifier was employed for metal transfer mode identification with notable accuracy. Figure 1.11 shows how arc voltage and current influence the transition between metal transfer modes. Broadly, free-flight transfers include globular and spray types:

- Globular Transfer: Occurs at lower current levels. Large droplets form at the wire tip and detach due to gravity. This mode is unstable and leads to higher spatter.
- Spray Transfer: Occurs at higher currents. Small, fine droplets are ejected from the wire tip due to electromagnetic forces. Spray transfer is further categorized into:
  - Projected Spray: Droplets detach near the solid–liquid interface. This transfer is stable and yields low spatter, making it ideal for industrial use.

- Streaming Spray: The molten electrode elongates before detachment. This is unstable and may lead to arc instability and inconsistent bead formation.

The choice of shielding gas also plays a vital role. For instance, with pure argon, spray transfer occurs at lower current levels and often manifests as streaming spray. With an argon–CO_2 mix, higher currents are required to achieve spray transfer. With pure CO_2, spray transfer is generally not achievable even at high currents. However, Kataoka et al. [25] demonstrated spray transfer in CO_2 using a specialized electrode and negative electrode polarity. Further research is essential to understand the interaction between the molten droplet and arc plasma, particularly under varying shielding environments and electrode geometries. In pulsed GMAW, metal transfer occurs through a controlled cycle of high and low currents. The peak current promotes droplet detachment in spray mode, while the background current sustains the arc with minimal heat input. This enables positional welding with high-quality droplet transfer and reduced heat distortion [11].

1.5 ADVANCES IN METAL TRANSFER MODES

With the advent of Industry 4.0, modern manufacturing lines increasingly rely on automation and real-time monitoring to enhance weld quality and productivity. However, literature on automated control of metal transfer modes remains limited, signaling a need for more research and innovation. GMAW is widely used in semi-automatic, mechanized, and fully automated welding systems. Future developments aim to improve arc stability, bead geometry control, metal deposition rate, and overall process efficiency. These advancements are crucial for the welding of diverse materials such as steels [26], aluminum, titanium, and magnesium alloys [27, 28]. Pulse parameters particularly peak current and pulse duration are pivotal in controlling metal transfer and weld bead quality. While most studies focus on peak conditions to optimize burn-off rate and droplet formation, background current conditions are often overlooked despite their potential influence on droplet dynamics and arc behavior. Expert welders are known to distinguish between metal transfer modes by the sound emitted during welding. Understanding this auditory feedback has led to psychoacoustic studies. For example, Tarn and Huissoon [29] evaluated how welders respond to arc sound cues to adjust welding parameters, providing insight into human-in-the-loop control strategies. Cullen et al. [24] further demonstrated the efficacy of signal-based classification. When using all four signals (current, voltage, acoustics, and gas flow), the metal transfer prediction rate improved from 80% to 99% using a feature selection algorithm. Notably, even with only the acoustic signal, prediction accuracy was 96% with the algorithm (vs. 81% without). This high accuracy highlights the potential of non-intrusive acoustic sensing for future smart welding systems capable of real-time adaptive feedback control. The move toward Industry 5.0 is accelerating the integration of AI and machine learning [30] into welding processes, shifting reliance from manual expertise to intelligent, data-driven systems. These technologies promise to revolutionize metal transfer monitoring and welding quality assurance.

1.6 DESIGN AND SYMBOLS

Proper weld design is fundamental to ensuring the structural integrity, mechanical strength, and functionality of welded assemblies. Weld design involves choosing the correct joint configuration, specifying weld size and type, and selecting suitable process parameters for a given application. In GMAW, the weld symbol typically includes a fillet triangle placed on the reference line of the engineering drawing, with "GMAW" indicated in the tail to denote the welding process. Additional information such as weld size, length, groove angle, and root opening may also be provided depending on the joint specification. Weld symbols serve as a universal language between design, fabrication, and quality control personnel. They encapsulate key instructions related to:

- Type of weld (e.g., fillet, groove, plug)
- Location (arrow side or other side)
- Size and length of weld
- Supplementary symbols (e.g., contour, finish method)

Figure 1.12 presents standardized welding design symbols, including both basic and supplementary types. These are based on international standards such as ISO 2553 and AWS A2.4, ensuring consistency across drawings and engineering

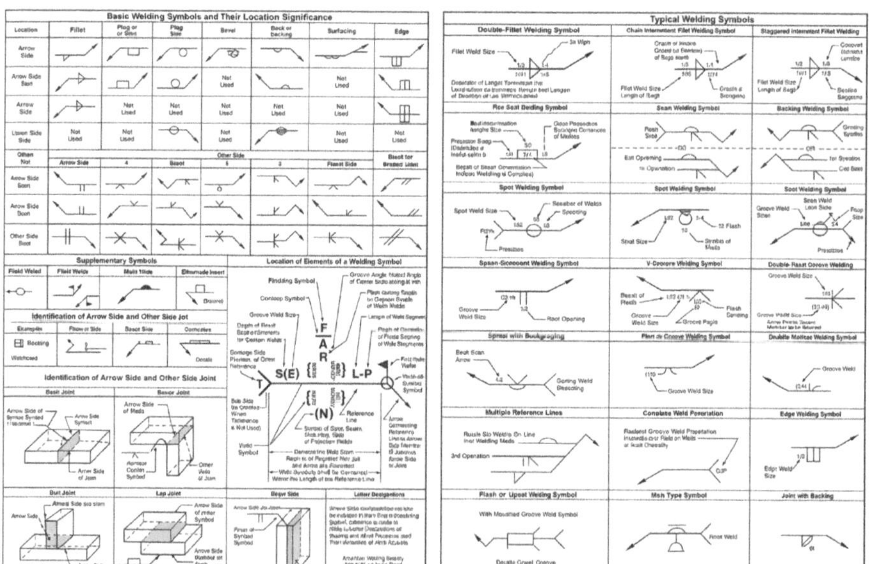

FIGURE 1.12 Welding design and symbols. (Courtesy: AWS [27]).

documentation. The weld size directly impacts the mechanical performance of the joint and must be specified carefully to meet strength requirements.

1.7 ADVANTAGES AND LIMITATIONS OF GMAW

Progressively, gas metal arc welding is employed for fabrication in many industries. The process is versatile since it can be functional for all-position welding. It can be easily automated and can easily be integrated into the robotized production centers. These advantageous features of this process have motivated various industries and many researchers to study the GMAW process in detail and explore its advantages across various fields.

1.7.1 Advantages

- GMAW can be used to weld all commercial metals and alloys
- The GMAW is an easily automated process
- No restriction of limited electrode length as we face in shielded metal arc welding (SMAW).
- Due to the gas shielding, no additional flux is required for the protection of the molten pool.
- GMAW can be used to weld in all positions.
- Higher deposition rates than shielded metal arc welding (SMAW).
- Due to the continuous electrode feeding mechanism, we can achieve higher welding speeds and higher filler metal deposition rates than shielded metal arc welding (SMAW).
- Due to the continuous electrode (wire feed) feed, longer welds can be deposited without intermediate stops and starts.
- Very little post-weld cleaning is required due to the absence of any heavy slag.
- Comparatively easy to learn and less skilled welders can also do the GMAW process.

Despite its wide application, the gas metal arc welding (GMAW) process has some limitations regarding the control of metal transfer. Although GMA welding was initially developed as a high deposition, high welding rate process facilitated by continuous wire feed and high welding currents, susceptibility to porosity and fusion defects limited its use to applications where weld quality was not of paramount importance.

1.7.2 Limitations

- The GMAW equipment is more complex, costlier, and less portable than that for SMAW.
- Due to the absence of flux covering, chances of faster cooling rates exist which may affect badly.

- Shielding of the molten weld pool is quite difficult during windy and drafty environments. High chances of porosity in the weld metal during windy environments.
- The GMAW process requires moderate cleaning of joints before welding.

1.8 CONCLUSIONS

The importance of gas metal arc welding (GMAW) in modern fabrication and manufacturing is underscored by its wide-ranging applications, including joining of dissimilar materials, automation integration, and additive manufacturing. This chapter has provided a comprehensive overview of GMAW—covering its fundamental principles, metal transfer modes, process parameters, recent advancements, and the increasing role of machine learning and acoustic signal monitoring in smart welding systems. The advantages of GMAW, such as high deposition rates, ease of automation, and adaptability to various metals and positions, make it a versatile and industrially viable technique. However, certain limitations—such as equipment complexity, sensitivity to environmental conditions, and challenges in transfer mode control—still pose practical concerns. With a focus on process efficiency, metallurgical integrity, and future innovation, GMAW continues to evolve as a key technology in Industry 4.0 and beyond [32]. The knowledge consolidated in this chapter sets a foundation for understanding the intricacies of GMAW and guides further research and industrial application.

The subsequent chapters of this book will explore GMAW in greater depth, particularly its integration into additive manufacturing workflows and advanced welding applications, emphasizing process optimization, material behavior, and state-of-the-art developments for next-generation manufacturing systems.

REFERENCES

1. R. W. Messler, *Principles of Welding.* Wiley, 1999. https://doi.org/10.1002/9783527617487.
2. S. Goka, S. N. Srirama, G. Divya Jyothi, S. Q. Moinuddin, and H. S. Shaik, "Welding-based 3D, 4D, 5D printing," in *Automation in Welding Industry.* Wiley, 2024, pp. 157–181. https://doi.org/10.1002/9781394172948.ch10.
3. AWS Standard, *Standard Welding Terms and Definitions*, AWS A3.0:2020, 2020, p. 157.
4. S. Goka, S. Q. Moinuddin, M. Cheepu, and A. K. Dewangan, "Welding practices in industry 5.0: Opportunities, challenges, and applications," in *Automation in Welding Industry.* Wiley, 2024, pp. 263–279. https://doi.org/10.1002/9781394172948.ch15.
5. H. Soltan and M. Omar, "A roadmap for selection of metal welding process: A review and proposals," *Weld World*, vol. 66, pp. 2639–2675, 2022. https://doi.org/10.1007/s40194-022-01379-1.
6. A. A. Y Al-Qenaei, "Fusion welding techniques," *J. Eng. Res. Appl.*, vol. 6, no. 2, pp. 78–83, 2016. www.ijera.com.
7. N. Stenbacka, I. Choquet, and K. Hurtig, "Review of arc efficiency values for gas tungsten arc welding," IIW Commission IV-XII-SG212, Intermediate Meeting, BAM, Berlin, Germany, April 18–20, 2012.

8. Md. Ibrahim Khan, *Welding Science and Technology*. New Age International, 2007.
9. J. Tapiola, "Cold metal transfer cladding of wear and corro- sion resistant coatings in engine applications," Jun. 2016. https://trepo.tuni.fi//handle/123456789/24583.
10. S. K. S. K. Khotiyan, "Investigation of microstructure and mechanical properties of TIG and MIG welding using aluminium alloy," Int. J. Educ. Sci. Res. Rev., vol. 1, no. 5, pp. 90–96, 2014. https://doi.org/10.9790/1684-130508121126.
11. P. K. Palani and N. Murugan, "Selection of parameters of pulsed current gas metal arc welding," *J. Mater. Process. Technol.*, vol. 172, no. 1, pp. 1–10, 2006. https://doi.org/10.1016/j.jmatprotec.2005.07.013.
12. P. Praveen, P. K. D. V. Yarlagadda, and M. J. Kang, "Advancements in pulse gas metal arc welding," *J. Mater. Process. Technol.*, vol. 164–165, pp. 1113–1119, 2005. https://doi.org/10.1016/j.jmatprotec.2005.02.100.
13. P. Kah and J. Martikainen, "Influence of shielding gases in the welding of metals," *Int. J. Adv. Manuf. Technol.*, vol. 64, no. 9–12, pp. 1411–1421, Feb. 2013. https://doi.org/10.1007/s00170-012-4111-6.
14. C. Li, C. Duan, Y. Qi, Y. Sun, and C. Ma, "Effect of shielding gas on MIG welding performance with austenitic wire," *Mater. Lett.*, vol. 339, p. 134118, May 2023. https://doi.org/10.1016/j.matlet.2023.134118.
15. G. Yang, F. Deng, S. Zhou, B. Wu, L. Qin, and J. Zheng, "Influence of shielding gas nitrogen content on the microstructure and mechanical properties of Cu-reinforced maraging steel fabricated by wire arc additive manufacturing," *Mater. Sci. Eng. A*, vol. 832, p. 142463, Jan. 2022. https://doi.org/10.1016/j.msea.2021.142463.
16. C. L. Jenney and A. O'Brien, eds., *Welding Handbook*, 9th ed., vols. 1–6. American Welding Society, 2001, pp. 89, 99–100, 119, 130–131, 140, 351.
17. TWI, "Power source characteristics," https://www.twi-global.com/technical-knowledge/job-knowledge/power-source-characteristics-121.
18. S. Kou, *Welding Metallurgy*, 2nd ed. Wiley-Interscience, 2003, pp. 37, 393.
19. W. D. Callister, Jr. and D. G. *Rethwisch, Materials Science and Engineering: An Introduction*, 8th ed. John Wiley & Sons, 2009, p. 344.
20. E. M. M. Puñales and S. C. A. Alfaro, "Stability on the GMAW process," 2021. https://doi.org/10.5772/intechopen.90386chopen.com/chapters/70416.
21. P. Kah, H. Latifi, R. Suoranta, J. Martikainen, and M. Pirinen, "Usability of arc types in industrial welding," *Int. J. Mech. Mater. Eng.*, vol. 9, no. 1, p. 15, Dec. 2014. https://doi.org/10.1186/s40712-014-0015-6.
22. P. Jiangang *et al.*, "Influence of arc mode on the microstructure and mechanical properties of 5356 aluminum alloy fabricated by wire arc additive manufacturing," *J. Mater. Res. Technol.*, vol. 20, pp. 1893–1907, 2022. https://doi.org/10.1016/j.jmrt.2022.08.005.
23. A. Scotti, V. Ponomarev, and W. Lucas, "A scientific application oriented classification for metal transfer modes in GMA welding," *J. Mater. Process. Technol.*, vol. 212, no. 6, pp. 1406–1413, Jun. 2012. https://doi.org/10.1016/j.jmatprotec.2012.01.021.
24. M. Cullen, S. Zhao, J. Ji, and X. Qiu, "Classification of transfer modes in gas metal arc welding using acoustic signal analysis," *Int. J. Adv. Manuf. Technol.*, vol. 115, no. 9–10, pp. 3089–3104, Aug. 2021. https://doi.org/10.1007/s00170-021-07305-x.
25. T. Kataoka, R. Ikeda, M. Ono, K. Yasuda, and Y. Hirata, "Effect of REM addition of wire on CO_2 gas shielded arc phenomenon," *Weld. Int.*, vol. 23, no. 7, pp. 517–522, 2009.
26. F. K. Tesfaye, "Parameter optimizations of GMAW process for dissimilar steel welding," *Int. J. Adv. Manuf. Technol.*, vol. 126, no. 9–10, pp. 4513–4520, Jun. 2023. https://doi.org/10.1007/s00170-023-11356-7.

27. S. Goka, M. Mallaiah, and M. J. Davidson, "Optimization of bead morphology for cold metal transfer wire arc additive manufacturing of AZ31 magnesium alloy wires," 2024, pp. 129–137. https://doi.org/10.1007/978-3-031-50240-8_27.
28. S. Goka, M. Manjaiah, and M. J. Davidson, "Study on mechanical, microstructural and corrosion analysis of wire arc additive manufactured AZ31 magnesium alloy," *Met. Mater. Int.*, 2024. https://doi.org/10.1007/s12540-024-01864-w.
29. J. Tarn and J. Huissoon, "Developing psycho-acoustic experiments in gas metal arc welding," IEEE International Conference Mechatronics and Automation, 2005, IEEE, pp. 1112–1117. https://doi.org/10.1109/ICMA.2005.1626707.
30. S. Goka, G. S. Narayana, G. Divya Jyothi, H. S. Shaik, and S. Q. Moinuddin, "AI and ML in welding technologies," in *Automation in Welding Industry*. Wiley, 2024, pp. 73–90. https://doi.org/10.1002/9781394172948.ch5.
31. "AWS welding chart," p. 7. https://doi.org/https://files-us-prod.cms.commerce.dynamics.com/cms/api/tflzsjpjnb/binary/MA2yQh.
32. S. Q. Moinuddin, S. H. Saheb, A. K. Dewangan, M. M. Cheepu, and S. Balamurugan (Eds.), Automation in the Welding Industry: Incorporating Artificial Intelligence, Machine Learning and Other Technologies. John Wiley & Sons, 2024.

2 Advances in Metal Transfer Modes

2.1 SHORT CIRCUIT TRANSFER (SCT)

Short circuit transfer is also called short arc and dip transfer. It is one of the most frequent and earliest standard modes of metal transfer in GMAW. During short circuit mode, the molten metal is transferred from the wire to the workpiece during welding and the repeated electrical short circuits are generated due to the molten droplets touching the molten weld pool. This mode results in the generation of low heat input and good control of the weld pool, as shown in Figure 2.1 [2].

2.2 ADVANCES IN SCT

Short circuit transfer was first identified and tended to be used in the 1950s during the development of the GMAW welding process. This mode was naturally identified during the experimental work conducted using low voltage and small diameters of filler wires in GMAW. This dip transfer behavior was further observed deeply and used with the introduction of different power sources with current and voltage control features. The working principle of this short circuit transfer is explained with the wire feeding continuously toward the work or weld pool; whenever the wire tip reaches the weld pool, a short circuit is created. This causes the current to spike, heating the wire and detaching the droplet due to the pinching effect. The detached molten droplet reaches the weld pool and the arc reignites [2]. This cycle repeats several times, such as 50–200 times in a second, as shown in Figure 2.2. This kind of metal transfer usually forms when the welding current ranges from 50 A to 200 A and the voltage from 16 V to 22 V. Wire diameter is critical for this, and it can be 0.8 to 1.2 mm under the direct current electrode positive polarity. Along with this, it is also important to understand that the effect of shielding gas has a strong influencing role. The shielding gas of 100% CO_2, 7–85% Ar, 15–25% CO_2 mixtures, or pure forms of CO_2 shielding gas has much effect on this kind of metal transfer mode. Another advantage of using CO_2 is that the penetration is deeper, but it generates more spatter [3]. However, less CO_2 mixture gas offers better arc stability and less spatter. In particular, it is also linked with the thickness of the material, and it is excellent for materials 1–6 mm thick, such as carbon steel, stainless steel, galvanized steel, and low alloy steels. In order to utilize this mode of metal transfer, it has been widely investigated and various developments in this type of metal transfer mode have been highlighted as follows.

DOI: 10.1201/9781003778844-2

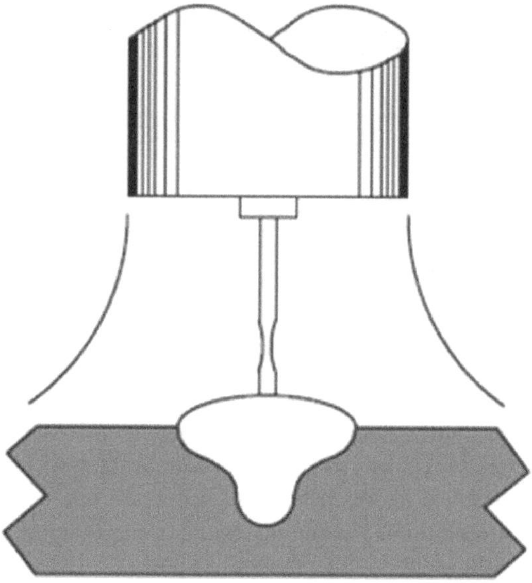

FIGURE 2.1 Schematic representation of short circuit transfer on GMAW [2].

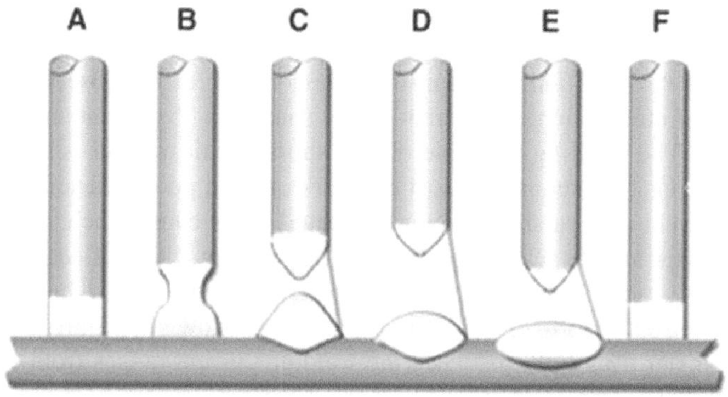

FIGURE 2.2 Short circuit cycle explains (A) wire and parent metal short circuit, (B) neck down, (C) wire isolation and the formation of the arc, (D) optimal welding arc and load voltage, (E) warmth of the arc, and (F) arc is off, repeat the cycle [2].

2.2.1 Surface Tension Transfer

Surface tension transfer (STT) is a kind of short circuit metal transfer controlled by waveform and is developed to enhance the arc stability and to reduce spatter, along with the precision welds for the open root and thin materials. This is an advanced short circuit metal transfer mode representing a technological leap with intelligent current modulation, as shown in Figure 2.3 [2]. Surface tension transfer was initially

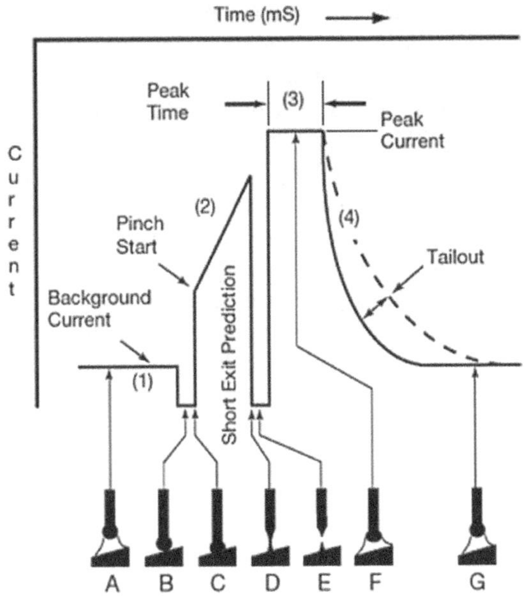

FIGURE 2.3 A schematic representation of STT current waveform: (A) wet-in (melting edge), (B) droplet separation, (C) pinch effect, (D) droplet separation, (E) upsurges the amperage, (F) peak time/current, and (G) tail-out [2].

developed and patented by one of the advanced welding research industries. The aim of this STT mode of metal transfer is to resolve the problems encountered by conventional short circuit transfer considering the reduction of spatter and improving arc stability, and integrating this system into the welding machines. Using this STT method, it is easy to maintain consistent penetration, controlled metal transfer, and stable arcs [4].

The principle of operation depends on the modulation of dynamic current and is synchronized with the metal transfer cycle. STT has three major phases: background current phase, pinch current phase, and tail-out/boost phase. This method appears usually at the welding current of 60 A to 200 A, voltage 18 V to 22 V, shielding gas of Ar + 10–15% CO_2 or Ar + 2% O_2. This process is most suitable for the welding of root passes in pipeline welding, thin materials, the automotive industry, and the fabrication of pressure vessels.

2.2.2 Cold Metal Transfer

Cold metal transfer (CMT) is an advanced gas metal arc welding process with the presence of low heat input and can be used for the precise metal transfer delivery in welding using the controlled wire motion from back and forth with synchronization with current pulses [5]. It was developed as an advanced version of the short circuit metal transfer process. The CMT process combines intelligent wire motion control with controlled current waveforms to produce smooth, spatter-free welds, as illustrated in Figure 2.4. The merit of this process is that it can weld thin sheets, dissimilar welds, and out-of-position joints.

Cold metal transfer was initially developed and patented by Fronius International GmbH, an Australian welding company, in the 2000s. The company aimed to develop this method to produce welds with less heat input to prevent burn-through and for precise metal transfer modes. In this CMT method, a synchronizing wire feed motion occurs with real-time current waveform control using a servo motor by retracting the wire during short circuiting. This system is fully controlled, and the cycle occurs 70–80 times per second. It operates in four stages: initially, the wire advances toward the weld pool by pushing the servo motor; when the wire reaches the weld pool, it creates a short [6]. Then in the third stage, the wire retracts and current drops. In the fourth stage, the arc re-establishes and the cycle repeats, as shown in Figure 2.5.

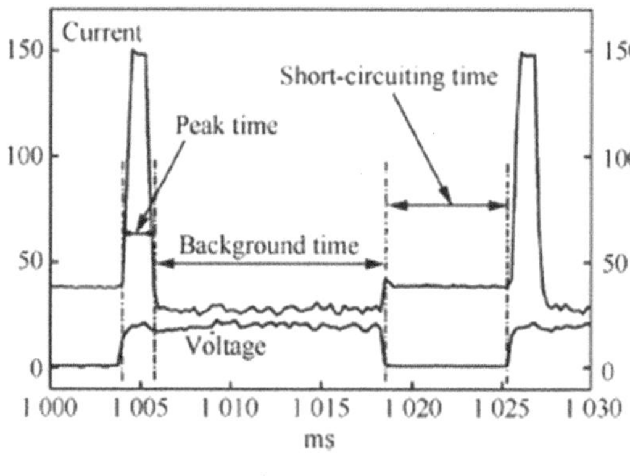

FIGURE 2.4 The cold metal transfer (CMT) cycle represents metal transfer mode [2].

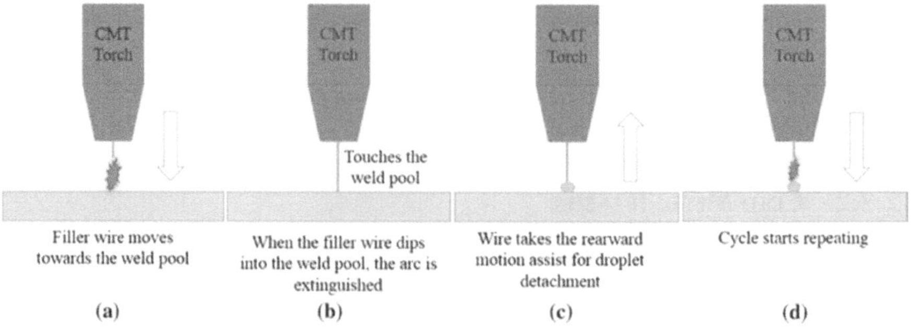

FIGURE 2.5 The metal transfer and movement of the wire electrode in the cold metal transfer (CMT) [7].

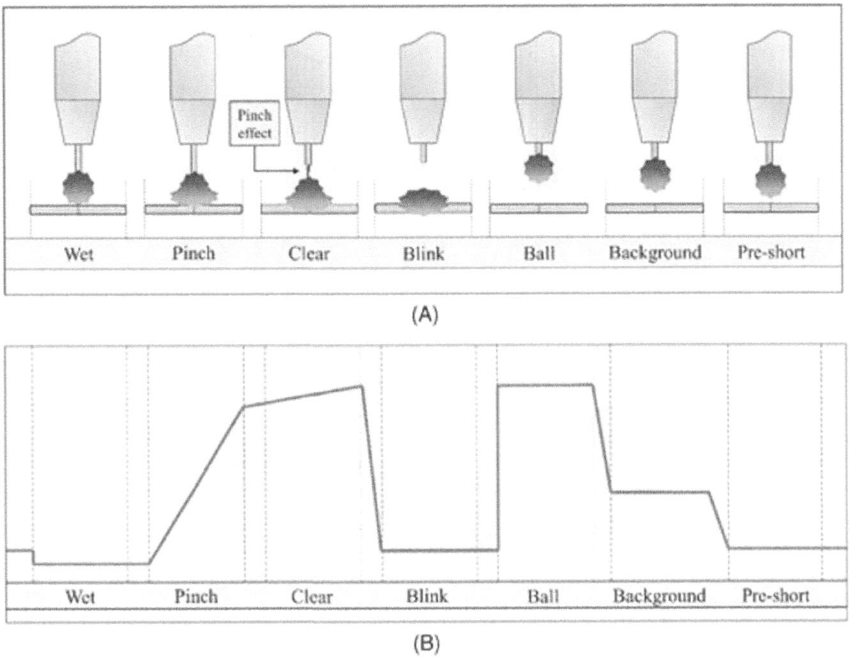

FIGURE 2.6 Metal transfer cycle for RMD: (A) various stages of metal transfer and (B) current waveform [2].

This method appears usually at the current range of 30–150 A and voltage 12–22 V. Wire diameters are 0.8–1.2 mm. The major advantages of this method are minimal distortion, reduced heat-affected zones (HAZ), and smooth beads.

2.2.3 Regulated Metal Deposition

Regulated metal deposition (RMD) welding was invented and patented by Miller Electric Mfg. In 2004 with the novel technique of the short-circuiting principle of metal transfer. RMD is recognized as one of the GAMW methods, a controlled short-circuiting operation, which guides the current by electronic means at every stage [8]. Computer program is needed to operate this and monitor the filler wire current at short circuiting. The company released that the RMS software program functions synergistically with an inverter-based arrangement coupled with closed loop feedback, and with a precision of 50 μs it can monitor and regulate the filler metal, as shown in Figure 2.6 [2].

2.3 GLOBULAR TRANSFER

Globular transfer is a kind of metal transfer that appears between the short circuit transfer and the spray transfer. It appears that, in general, the welding current,

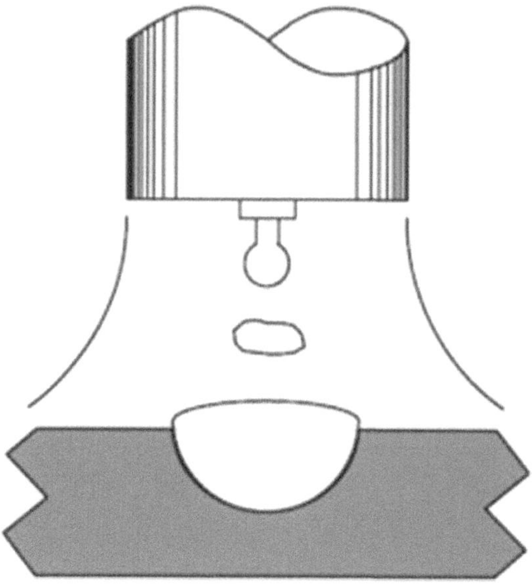

FIGURE 2.7 Globular metal transfer [2].

voltage, and wire feed speed are excessive compared to the short circuit transfer. It can occur at low welding current and higher voltages with the shielding gases of CO_2 and helium [9]. In this metal transfer, balls of molten filler wire are sprinkled all over the welding arc as a result of gravitational force, as shown in Figure 2.7.

2.4 SPRAY TRANSFER

Spray transfer is a smooth and clean process, where spatter almost does not appear during welding. In this process, a stream of spray with small molten droplets from the wire transfers to the weld pool under the influence of strong electromagnetic forces, as shown in Figure 2.8 [2, 10].

2.5 ADVANCES IN SPRAY TRANSFER

2.5.1 Pulsed Spray

Pulse spray transfer, also referred to as pulsed MIG welding, is a modified spray transfer process in which the welding power source rapidly alternates between a high peak current and a low background current. During the peak current phase, a molten droplet is pinched off from the electrode tip and propelled toward the weld pool, ensuring good fusion. The background current maintains the arc but is too low to allow metal transfer, as shown in Figure 2.9. This controlled droplet formation reduces spatter and allows welding in all positions, making it especially suitable for thin materials, aluminum, and stainless steel.

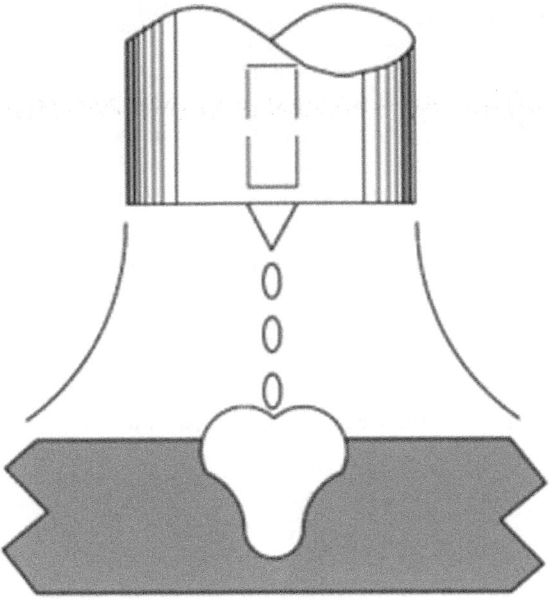

FIGURE 2.8 Spray metal transfer [2].

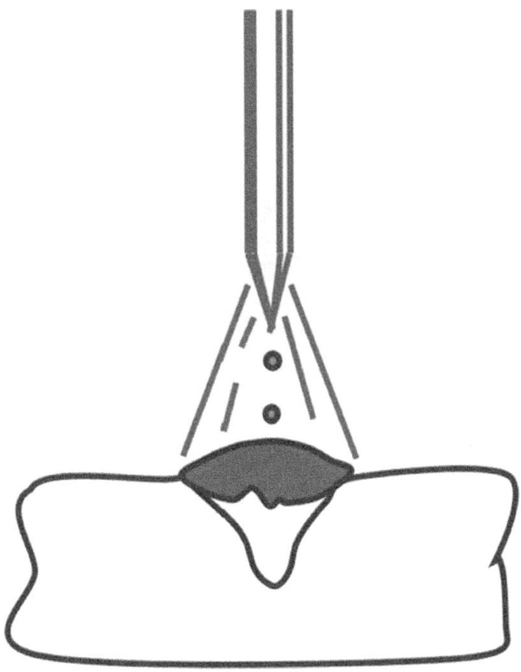

FIGURE 2.9 Pulsed spray metal transfer. (Courtesy: Miller Welds).

2.5.2 Rotational Spray

Rotating spray transfer occurs at very high welding currents when the liquid metal at the electrode tip rotates as it transfers into the weld pool. The rotational motion of droplets is influenced by electromagnetic forces in the arc, resulting in enhanced wetting and spreading of the weld pool, as shown in Figure 2.10. Although it can produce slightly higher spatter than axial spray, this mode improves bead uniformity and is particularly useful in aluminum welding or for complex joint geometries. It is considered stable and efficient for high-current applications.

2.5.3 Axial Spray

Axial spray transfer is characterized by a continuous, straight-line transfer of fine molten droplets from the electrode tip directly into the weld pool. This mode

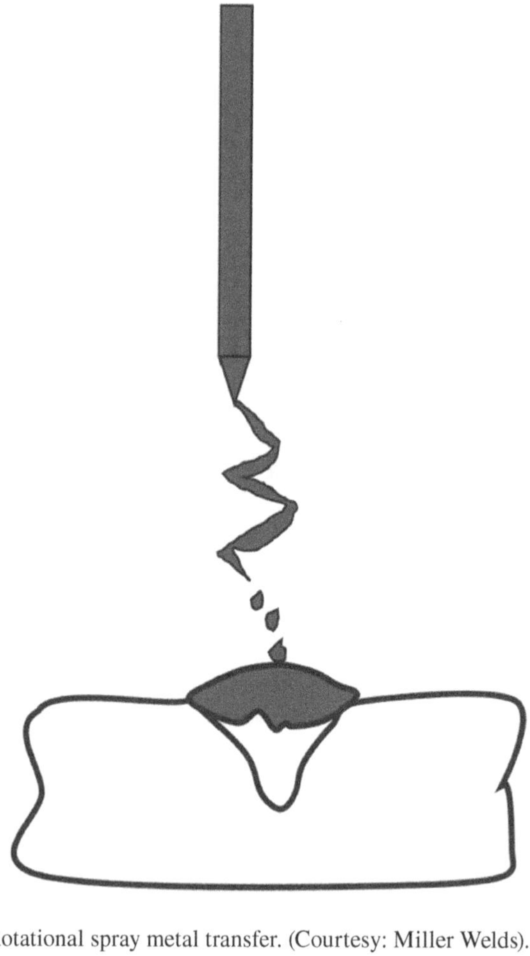

FIGURE 2.10 Rotational spray metal transfer. (Courtesy: Miller Welds).

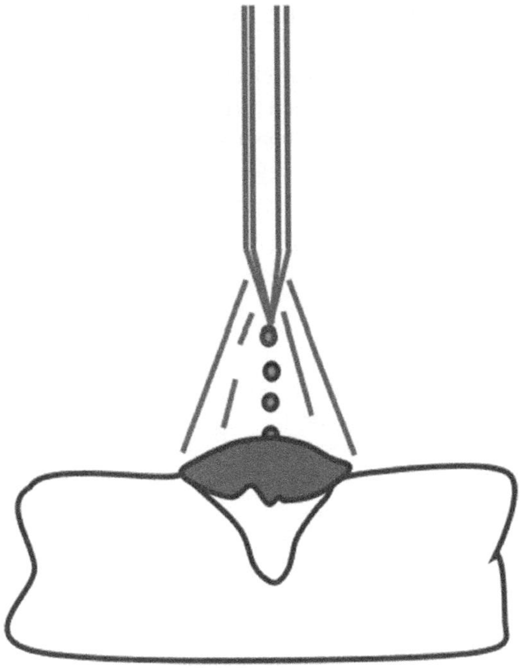

FIGURE 2.11 Axial spray metal transfer. (Courtesy: Miller Welds).

typically requires higher welding current and voltage, producing a stable arc with minimal spatter. It is ideal for flat and horizontal welding positions and is commonly applied in thick-section steel welding, such as shipbuilding and heavy fabrication. The straight, axial droplet path allows for high deposition rates and excellent bead quality, as shown in Figure 2.11.

2.6 CONCLUSION

Understanding and selecting the appropriate metal transfer mode in GMAW is essential for achieving optimal weld quality, process stability, and productivity. This chapter has highlighted both the conventional and advanced metal transfer modes—including short circuit, globular, and spray transfer, along with their enhanced versions such as STT, CMT, and RMD. Each mode brings unique advantages depending on the application, material thickness, joint position, and desired mechanical properties.

Innovative transfer methods like STT, CMT, and RMD represent a shift toward precision-controlled and low-heat input welding, enabling applications such as thin sheet welding, open root passes, dissimilar material joining, and out-of-position welding. Simultaneously, developments in spray transfer, including pulsed, rotational, and axial spray modes, have significantly improved deposition rates, arc stability, and spatter control—especially valuable in automated and additive manufacturing environments.

As GMAW continues to evolve, the selection and control of metal transfer modes are becoming increasingly sophisticated, driven by advancements in waveform control, digital power sources, and intelligent systems. A thorough understanding of these transfer mechanisms empowers welding engineers and researchers to tailor the welding process for diverse industrial needs, making metal transfer mode knowledge not only foundational but also a gateway to high-performance and adaptive welding technologies.

REFERENCES

1. Y. S. Kim and T. W. Eagar, "Analysis of metal transfer in gas metal arc welding," Weld. J.-New York, vol. 72, pp. 269-s, 1993.
2. V. D. Prajapati, J. J. Vora, and K. Abhishek, "Advances in gas metal arc welding process: Modifications in short-circuiting transfer mode," Adv. Weld. Deform., pp. 67–104, 2021.
3. S. Rhee and E. Kannatey-Asibu, "Observation of metal transfer during gas metal arc welding," Weld. J.-New York, vol. 71, pp. 381-s, 1992.
4. J. C. Dutra and R. G. de Silva, "MIG/MAG—short-circuit metal transfer—control of the current wave form—fundamentals of the CCC and STT systems," Weld. Int., vol. 23, no. 3, pp. 181–185, 2009.
5. C. G. Pickin, S. W. Williams, and M. Lunt, "Characterisation of the cold metal transfer (CMT) process and its application for low dilution cladding," J. Mater. Process. Technol., vol. 211, no. 3, pp. 496–502, 2011.
6. S. Venukumar, M. Cheepu, T. Vijaya Babu, and D. Venkateswarlu, "Cold metal transfer (CMT) welding of dissimilar materials: an overview," Mater. Sci. Forum, vol. 969, pp. 685–690. Trans Tech Publications Ltd, 2019.
7. D. Srinivasan, P. Sevvel, I. J. Solomon, and P. Tanushkumaar, "A review on Cold Metal Transfer (CMT) technology of welding," Mater. Today: Proc., vol. 64, pp. 108–115, 2022.
8. S. Das, J. J. Vora, V. Patel, W. Li, J. Andersson, D. Y. Pimenov, K. Giasin, and S. Wojciechowski, "Experimental investigation on welding of 2.25 Cr-1.0 Mo steel with Regulated Metal Deposition and GMAW technique incorporating metal-cored wires," J. Mater. Res. Technol., vol. 15, pp. 1007–1016, 2021.
9. J. J. Lowke, "Globular and spray transfer in MIG welding," Weld. World, vol. 55, pp. 19–23, 2011.
10. S. Rhee and E. Kannatey-Asibu, "Observation of metal transfer during gas metal arc welding," Weld. J.-New York, vol. 71, pp. 381-s, 1992.

3 Advances in Gas Metal Arc Welding

3.1 BACKGROUND

Fusion welding is the most popular type of welding in which an electric arc is created between an electrode and a workpiece. The substance and electrode wire are melted using the resulting arc heat. Due to their greater deposition rates and cost-effectiveness, gas metal arc welding (GMAW) and submerged arc welding (SAW) are two of the most often employed arc-welding techniques in industries. The GMAW method has seen tremendous advancements in the recent few decades in this competitive universe towards an automated industrial revolution [1]. Compared to single-wire welding, the requirements for deposition and production rates at the shop floor are more significant. Research is being done on new and current materials and technologies to overcome these constraints. Among these methods, the multiple-wire welding process was developed to boost output. This method is useful in situations when a fast rate of deposition without deep penetration is required. Examples include distinct butt and fillet welds where fit-up or melt-through is crucial, heavy corner welds, and surface deposits. Twin-wire welding was the main use at first. Later, as power sources and torch configurations (triple, four, and six wires) advanced, multiple wires were linked to one or more power sources. Ashton (1954) was the first to suggest the concept of multiple-wire welding in the form of twin-wire welding [2]. The older works of the 1950s and 1960s [3, 4] contain references to multi-wire welding. With the process introduction to the factory floor, interest in this technology began to rise in the 1980s. Following that, several advancements have been documented. The primary focus of the early research investigations was process setup and feasibility. Lassaline et al. [5] first utilized the multiple-wire concept in the GMAW process in the late 20th century. Subsequent reports also included twin and tandem wires with the GMAW. The advent of sophisticated power sources that offer fine control and modulation of the welding arc has led to an increase in research interest in multiple-wire GMAW throughout the past ten years [6, 7]. Owing to the flexibility of the GMAW process, it has been applied to a broad range of materials, including steel, aluminum, titanium alloys, and advanced materials like magnesium alloys [8–10]. Thus, the method has extended its services through 3D, 4D, and 5D printing [11, 12]. It is anticipated that multiple-wire welding will have a tremendous influence soon, requiring the consolidation of current technology and emerging technologies. The multiple-wire welding technique is briefly discussed in the next part, along with current advances, to help you comprehend the procedure. The chapter comprehensively

DOI: 10.1201/9781003778844-3

overviews technological advancements in multiple-wire gas metal arc welding applications. It also addresses major challenges, such as mutual arc interactions affecting arc stability, and discusses enhancements in process variants to mitigate these issues. Methods for controlling and measuring arc stability in multiple-wire welding are explored, along with future research directions in the field.

3.2 CLASSIFICATION OF MULTIPLE-WIRE WELDING

Given the noteworthy developments in the past few years, a complete classification of multiple-wire welding based on the torch, contact tube, power sources, polarity, and welding directions, as shown in Figure 3.1, summarizes the classification of whole categories of multiple-wire processes.

The multiple-wire welding extends conventional single-wire welding by feeding more than one electrode through one or more contact tubes. This technique increases the deposition rate, making it suitable for a wide range of applications in industries such as pressure vessels, shipbuilding, and pipelines. Various multiple-wire welding

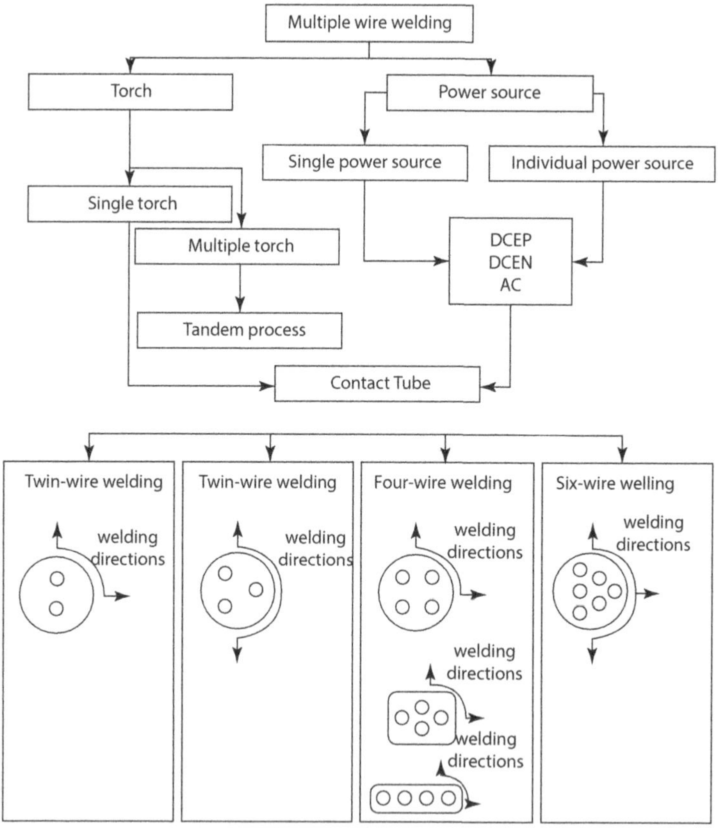

FIGURE 3.1 Classification of multiple-wire welding processes [13].

variants include tandem, transferred-ionized molten energy (TIME)-twin, double electrode, indirect arc types, cable-wire type welding, and metal-powder-assisted welding processes [14].

3.3 TWIN/TANDEM GMAW

Two-energy source welding has recently acquired much attention as an alternative approach to challenging welding defects. Such a method has more flexibility in controlling the heat input, thus being able to control the weld melt. A two-energy source welding system generally consists of a leading and a trailing energy source, as in the case of tandem laser beams, laser–arc, gas tungsten arc, and gas metal arc (GTA-GMA), and tandem GMA hybrid welding [15].

Tandem gas metal arc welding is a modified method of the GMAW process where two electrodes are simultaneously used to perform the welding operation. This process involves the deposition of two separate and independent arcs in the same weld pool. Tandem GMAW is known for its high welding speeds, increased deposition rates, and deep penetration capabilities. Additionally, it is beneficial for welding thick materials, as it can provide high-quality welds with fewer passes. By utilizing two electrodes, this process allows for improved control over the heat input and weld bead profile, resulting in enhanced productivity and quality in welding applications. Nevertheless, twin-wire welding-based additive manufacturing (TWAM) is a unique process that uses gas metal arc welding-based twin-wire weld-deposition to create functionally gradient materials with varying hardness [16, 17]. The process involves variation in positional direction as shown in Figure 3.2 . Thus, the obtained *V–I* characteristics are depicted in Figure 3.3.

In the last decade, the twin-wire GMAW process improved in use due to developments in technology, enabling optimum torch design and precise control over the welding system. In twin-wire GMAW, there is a considerable mutual impact between the arcs [19].

Figure 3.4 depicts the interaction between twin-wire electrodes. It's a distinct process due to understanding that the coalescence phenomenon of the weld joint is typical and complex. Electromagnetic interactions (EMI) and self-induced magnetic

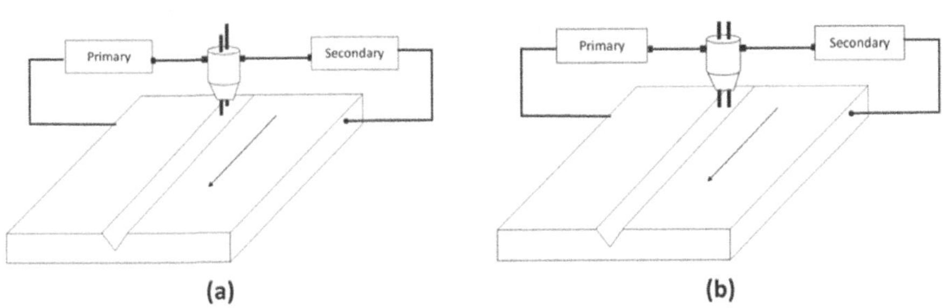

FIGURE 3.2 Twin-wire welding process: (a) tandem wire position (wires one after another) and (b) transverse wire position (wires one beside other) [18].

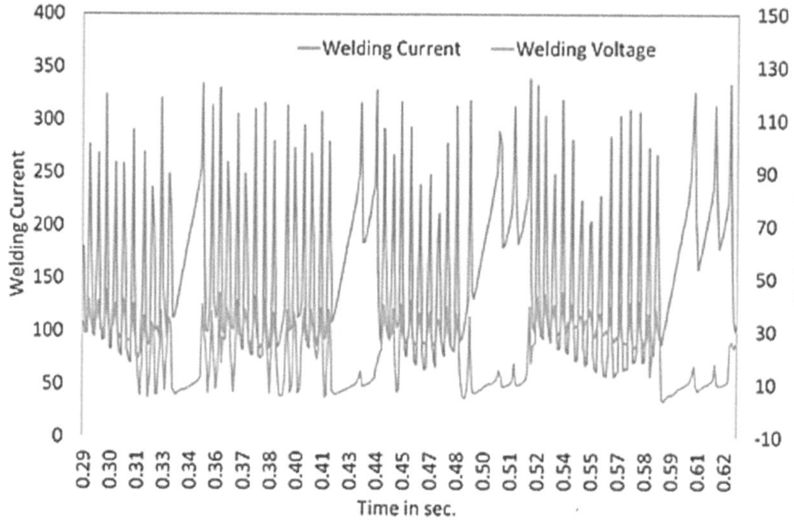

FIGURE 3.3 A typical current and voltage signal in the pulsed GMAW process [18].

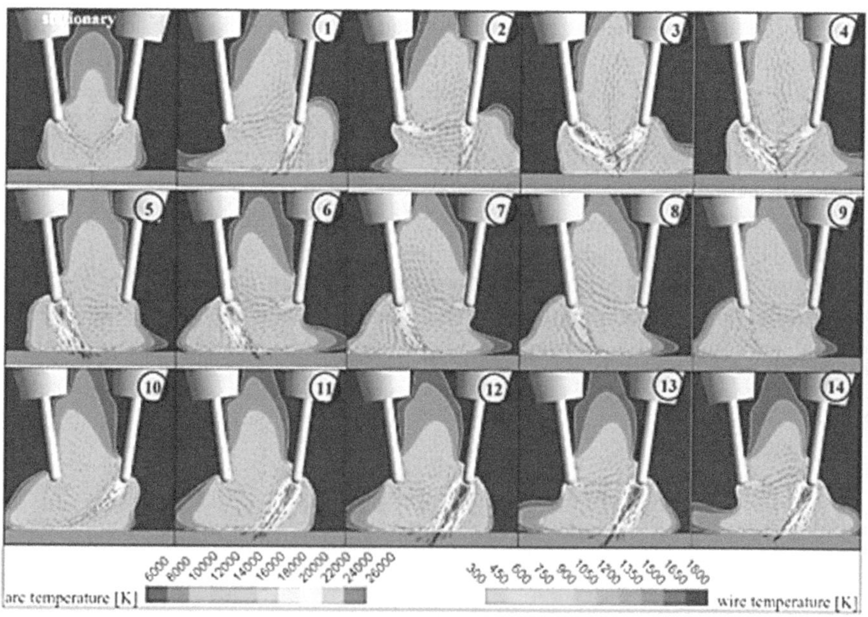

FIGURE 3.4 Interaction between twin-wire electrodes [20].

fields are the cause of this arc interaction. The arc interactions cause the weld pool to oscillate, which leads to arc deflection. This can cause the arc column to be non-perpendicular to the surface of the weldments and have poor mechanical properties. In addition, the arc interactions lead to welding process instability, which in turn affects the geometry of the weld bead. Transferred ionized molten energy (TIME), which enables the creation of a steady electric arc with a desired metal transfer at both wires, is the primary characteristic of this technique. This procedure, in contrast to traditional GMAW/SAW, makes use of a proprietary four-component gas combination ($Ar/He/CO_2/O_2$) and a customized nozzle that stabilizes the process by controlling metal transfers in all places. Larger electrode extensions are made possible by this nozzle, which raises the deposition speeds. Tandem welding is comparable to Fronius International GmbH's TIME-Twin.

Multiple-wire welding techniques, such as twin-wire gas metal arc welding (GMAW) and tandem submerged arc welding (T-SAW), have garnered significant academic interest. Numerous studies focus on optimizing welding parameters, heat inputs, and wire feed rates in twin-wire GMAW to enhance productivity while maintaining high weld quality. Additionally, extensive research on the interaction of electrical parameters with metal transfer modes in twin-arc integrated cold wire hybrid welding highlights the ongoing effort to improve welding efficiency [21]. The process, namely multiple wire arc additive manufacturing (MWAAM), emerged from the fundamental ideas of wire arc additive manufacturing (WAAM) [22]. Combining several wire feedstocks into a single manufacturing setup expands the possibilities of conventional WAAM. This kind of integration makes it possible to build components with material characteristics that are spatially unique, giving designers and engineers more options to optimize component attributes according to specific local requirements. The effective adoption of MWAAM requires addressing some inherent problems and limitations, despite its many benefits [23].

3.4 TRIPLE-WIRE GMAW

In contrast to the traditional twin-wire GMAW procedure, the triple/tri-arc twin-wire gas metal arc welding (GMAW) is a unique twin-wire welding technique that separates heat input and deposition rate by forming an inter-electrode arc between the two wire ends known as the middle-arc (M-arc). The M-arc boosts the current to melt one welding wire while preheating the second wire without passing through the workpiece, which is how the tri-arc twin-wire welding technique improves deposition rate and reduces heat input. Figure 3.5 depicts the power source used for the welding process. Also, it is represented clearly how the current flows through the loop of the system. Further, the arc shape and dynamic behavior of the process were identified to reveal the decoupling properties of this unique welding process more clearly.

Figure 3.6 represents the schematic of the theoretical waveform of the acquired current used for the tri-arc GMAW process. Here, the times "b" and "c" represent the time corresponding to Figure 3.5. Three operating modes for tri-arc twin-wire

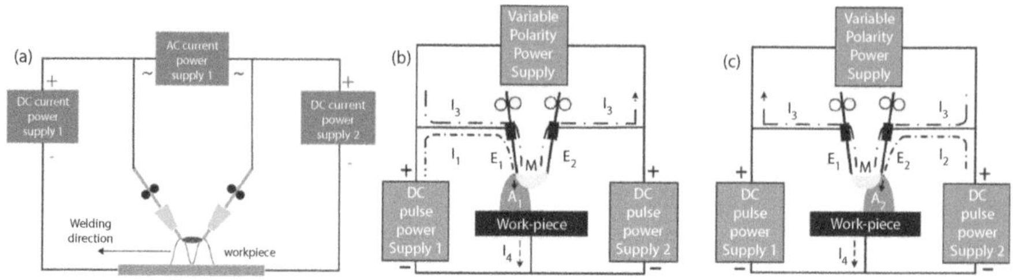

FIGURE 3.5 Schematic of power source used for the tri-arc welding: (a) whole composition [20] and (b, c) process decomposition [25].

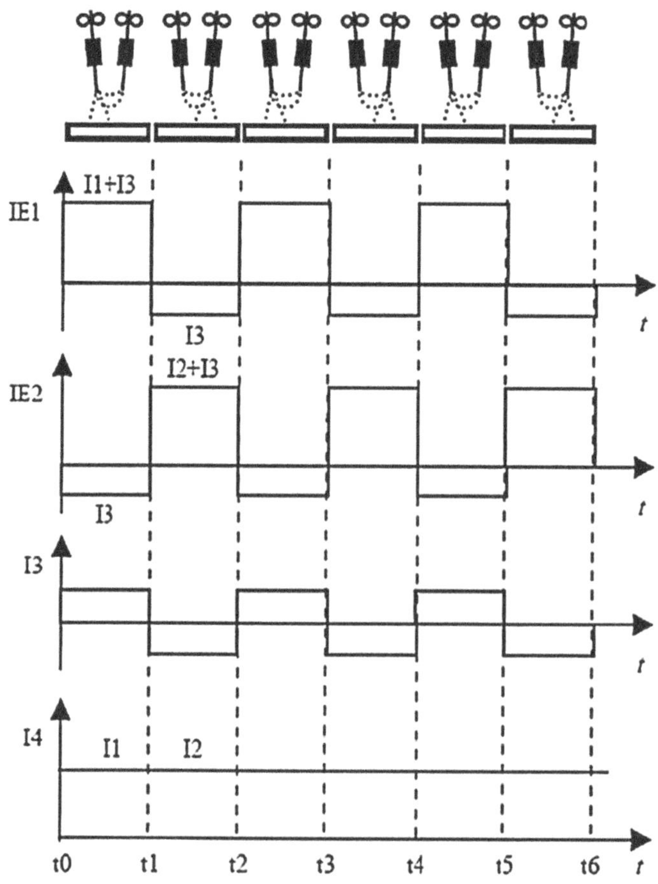

FIGURE 3.6 Schematic diagram of the waveform of the acquired current [25].

welding include single-wire single-arc, twin-wire dynamic double-arc, and twin-wire dynamic three-arc, according to an examination of the experimental data.

In the twin-wire dynamic three-arc welding process, at least two arcs dynamically produce two distinct arc forms and their corresponding mirror image shapes, a "μ" shape and a "Γ" shape. Furthermore, the method for creating the "Γ"-shaped arc is explained. The authors recorded welding current signals and dynamic arc morphologies using a synchronized data-collecting system and a high-speed camera.

3.5 FOUR/SIX-WIRE GMAW

Multiple-wire welding is an advancement over traditional single-wire welding in which numerous electrodes are supplied through one or more contact tubes. The procedure may be applied to a broad range of sectors, including pressure vessels, shipyards, pipelines, etc., since the multiple wires increase the deposition rate. The three pairs of wires in the triangle configuration, as represented in Figure 3.1, engage with one another similarly to twin-wire welding. Together, the three interactions produce a rotating arc that is oriented toward the center and aids in the process of achieving weld penetration.

As shown in Figure 3.1, there are several ways to organize the wires in four- and six-wire systems. At the shop floor level, the four-wire and six-wire welding procedures have not yet been developed. Two-wire arrangement systems hence dominate advancements in multiple-wire technologies. Twin-wire is the most utilized variation of this arrangement for overlays, whereas tandem is the most used type for welding.

3.6 CABLE WIRE GMAW

With a broad variety of market opportunities, cable wire arc welding is a new welding technique that boasts great efficiency, energy savings, simple equipment, welding automation, and good welding quality. This unique welding technique is called the cable-type welding wire (CWW) arc welding process, in which the wire is employed as a consumable electrode. There are seven welding wires in the CWW; the central wire has one welding wire, and the remaining six are evenly spaced around it. There is just one wire feeder, one welding torch, and one welding power supply utilized in the welding operation. The seven wires melting at the same time allow for a significant increase in the deposition rate. The quality of the welding is enhanced when a self-rotating arc is formed by the six periphery wires revolving around the core wire. Under the same circumstances, the deposition rate of CWW CO_2 welding increased by 40% over single-wire CO_2 welding [26].

The center wire, also known as the flux-cored wire or solid wire, is located in the middle of the CWW type of rotational stranded wire. As shown in Figure 3.7, the peripheral wires form a large diameter CWW by twisting around the core wire. The wire feeder feeds the CWW constantly, causing the seven branch wires to melt at the same time. The six peripheral wires then rotate around the central wire to produce a self-rotating arc. Some drawbacks of conventional welding wire, including its small diameter, coil challenges when expanding the wire diameter, and low stiffness, are

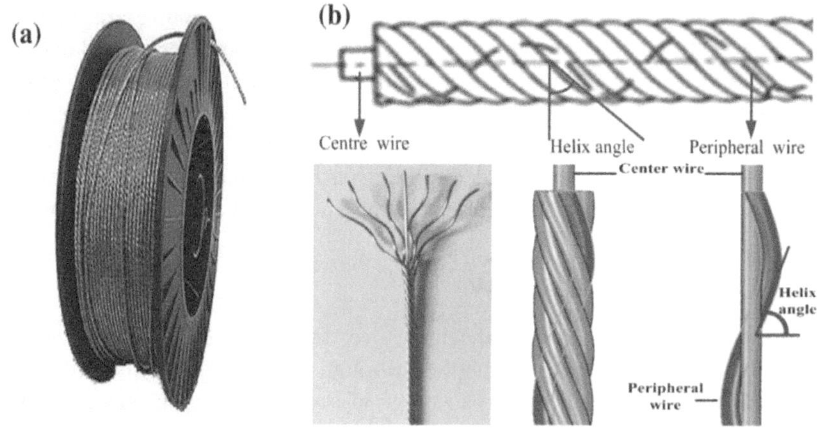

FIGURE 3.7 Schematic diagrams of cable welding wire: (a) as received type, (b) longitudinal view, (c) solid center wire type, and (d) center flux cored wire type [27].

addressed by CWW. By modifying one or more peripheral wire alloy components, the CWW can create a novel method for enhancing the qualities of welded joints.

One of the main areas of research in cable wire arc welding is the types and processes of twisting. The primary categories of cable type wires (a) 1*7, (b) 7*7, (c) 6*7+1, and (d) 1*19 are used for CWW arc welding. Solid and flux-cored wires are examples of branch wires. Solid and flux-cored wires can easily collide, which enhances the welding characteristics. The most popular cable wire is the 1*7 type [26].

3.6.1 Principle of CWW

Figure 3.8 illustrates the CWW welding system's equivalent circuit. The direction of the current flow during the CWW welding process is from the seven wires to the base metal. The current forms after the wire comes into contact with the metal. Both the melting of the CWW and the heating and melting of the base metal are the

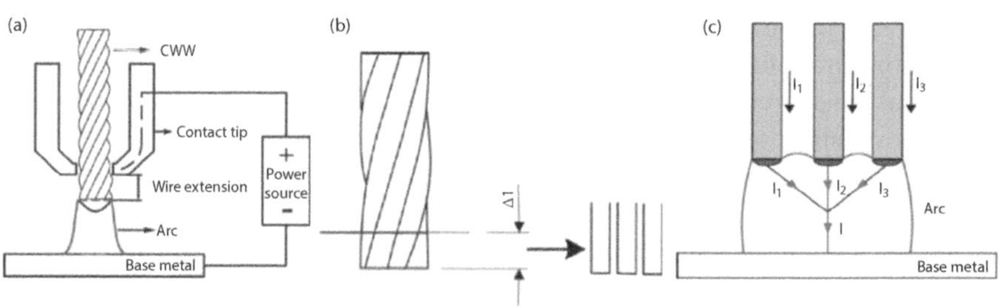

FIGURE 3.8 Schematic diagram CWW CO_2 welding: (a) equivalent circuit, (b) cable wire tip, and (c) distribution of current [28].

two primary uses for the energy produced. Figure 3.8 shows an examination of the stresses and current flowing through the tip of the CWW, which is thought of as a group of parallel wires. The current distribution during CO_2 welding with CWW is seen in Figure 3.8(c).

The arc shape is formed by current passing from the wire to the base metal. We call this phenomenon arc coupling when the currents couple together to produce a single current channel. These three currents, which correspond to the separate wires, add up to the total welding current on the 11′ axis.

3.6.2 The Mechanism of Formation of the Arc Shape and Weld Seam

Figures 3.9 and 3.10 depict the arc behavior observed during CWW CO_2 welding. In contrast to the arc diameter fluctuation seen in a bell-shaped arc, it is evident that the arc diameter decreases along the direction of the arc length. Every wire end has a single brilliant anode spot, which causes each wire to progressively melt due to the anode spot's activity. Furthermore, the anode spot rotates at the same rate as the twist wires in the CWW, moving together with the wire. A stable arc is produced as a result of the anode spot driving the arc rotation. Each wire melts and generates a droplet; the arc force then attracts the droplets from each wire to form a single, massive droplet.

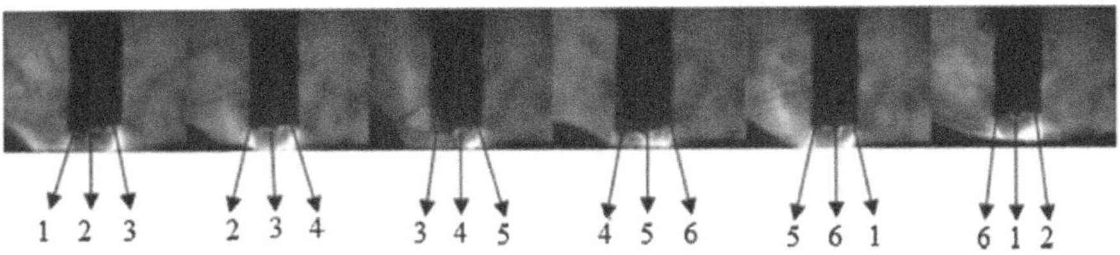

FIGURE 3.9 The shape of the arc in cable weld wire arc welding [28].

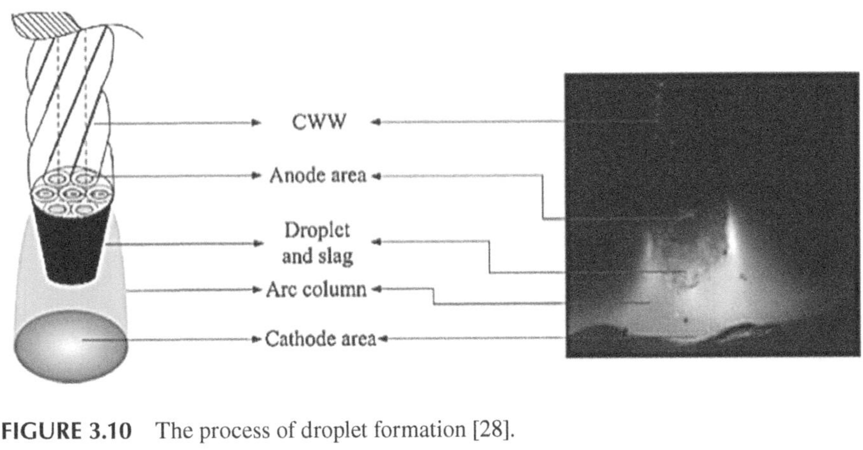

FIGURE 3.10 The process of droplet formation [28].

The mechanism of droplet generation is shown in Figure 3.10. As the wire is fed into and revolves around the center wire of the CWW, the tips of each wire melt. The distinct twisting properties of the CWW being employed determine the rotational features of the one spinning arc that swiftly unites the arcs from the separate wires. The individual wire tips are forced to move in the CWW's winding direction by the mechanical feeding process and wire melting, which causes the wire tips to rotate synchronously with the coupling arc.

The electromagnetic forces operating on three wires along the 11′ axis are schematically shown in Figure 3.11. The arc will be deflected by attraction forces that arise between the three wires on the 11′ axis since their currents run in the same direction.

Figure 3.12(a) illustrates the arc forms for both single-wire arc welding and CWW CO_2 welding. The arc diameter in CWW CO_2 welding is comparable to the CWW diameter, and the weld pool adopts a concave form as opposed to single-wire welding. In CWW CO_2 welding, the arc is focused in the center and has a beam form, which is compatible with the deflection length model previously discussed. This is in contrast to the arc shape in single-wire welding. Figure 3.12(b) depicts the stability

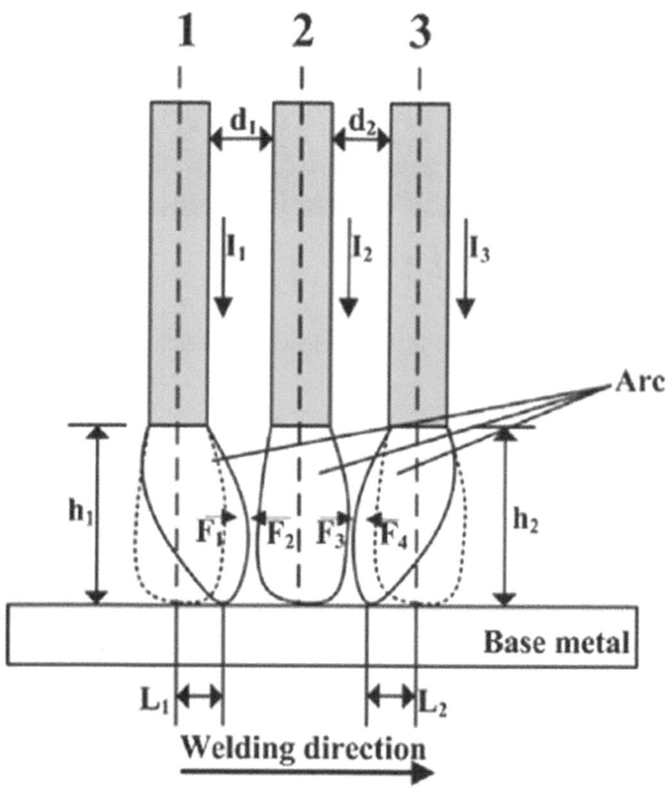

FIGURE 3.11 Electromagnetic force acting on three wires [28].

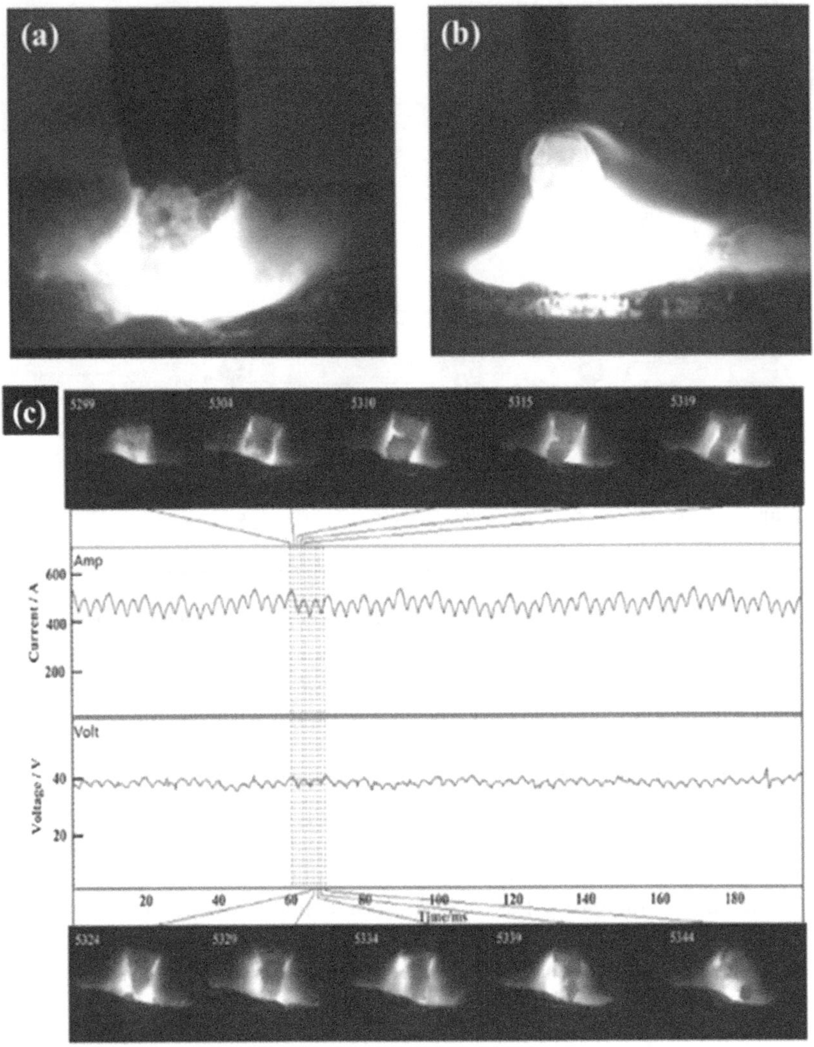

FIGURE 3.12 Arc characteristics in CWW welding: (a) arc shape in cable wire CO_2 welding, (b) arc shape in single wire welding, and (c) arc stability in cable wire CO_2 welding [28].

formation of the arc in CWW CO_2 welding. Because of the symmetrical forces acting on the arc during the welding process, the arc keeps its beam form while the wire spins and the weld pool doesn't stir. The waveforms of voltage and current recorded throughout the welding operation also show the stability of the CWW CO_2 welding arc. In line with the arc stability shown by the high-speed camera system, the current and voltage change steadily throughout droplet generation and transmission. The arc rotation and motion trajectory in CWW CO_2 welding are schematically illustrated in Figure 3.13. In a circle with six symmetrical places ranging from t1 to t6, the outer

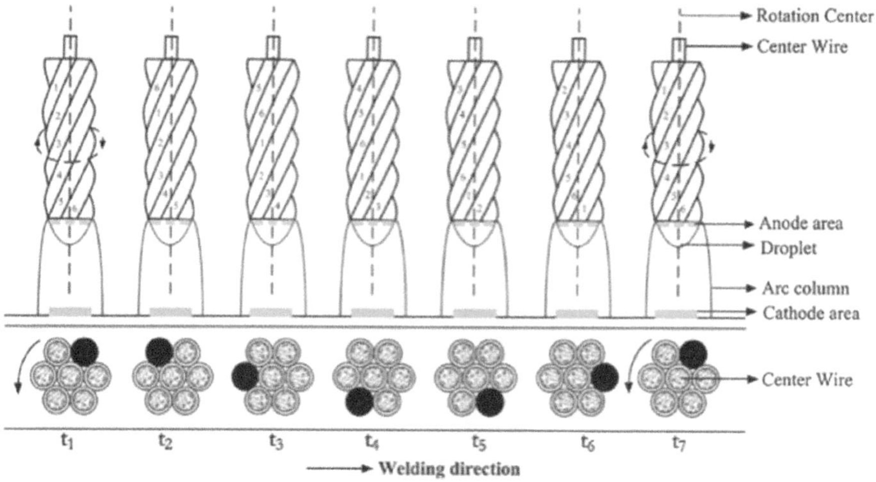

FIGURE 3.13 Rotation of arc and trajectory of arc motion [28].

wires revolve around the central wire. The droplets congregate under the arc coupling effect, and as soon as the arc is formed, the wires melt. The arc rotates as the motor feeds the CWW because each wire's anode location rotates. Maintaining the initial beam form of the arc, the accompanying forces revolve around the wire's core. Following its descent from the wire, the droplet on the wire will have some velocity and spin with the wire.

Many issues, including unequal tension, loose wires, varying wire diameters, and residual stress, can result in unsmooth wire feeding during the CWW twisting process, which can compromise the stability of the welding process. Clear modulations of various wires under the specific operating conditions for CO_2 gas were observed.

As the current increases, the droplet transfer mode switches between rejected, globular, and projected transfers. The various forces acting on the droplet are what give rise to various properties of the droplet transfer modes. Among the three droplet transfer modalities, a rejected droplet has the biggest size due to the combined forces that resist the droplet during the transfer process. The droplet transfer process in CWW GMAW for various welding conditions is captured by the high-speed digital camera [26]. As the weld parameters are increased, the result for CWW GMAW indicates three types of droplet transfers: repelled, globular, and projected. For a welding current of 450 A and a welding voltage of 26 V, the arc shape and droplet picture for the rejected droplet transfer. Bell-shaped is the arc. Droplets are assisted in transferring off the wire by the spinning force of the CWW, which causes the droplet to rotate during the process. The droplet size is 1.5 times the diameter of the CWW, and the droplet transfer frequency is around 20 Hz. For a welding voltage of 28 V and a welding current of 500 A, the arc shape and droplet image of the globular droplet transfer. Because of the arc's compression, the arc is also bell-shaped but smaller. The droplet size is two-thirds the diameter of the CWW, the droplet transfer frequency is around 40 Hz, and the droplet transfer mechanism is steady with minimum spatter. The anticipated droplet transfer with a welding current of 550 A and a welding voltage of 31 V is depicted in, together with the arc shape and droplet picture. With increasing welding conditions, the arc gets smaller and eventually takes on the shape of a bundle arc. Rising downward pressures result in a reduction in droplet size. The droplet size is half the diameter of the CWW, and the droplet transfer frequency is around 100 Hz. As welding parameters increase, the anticipated droplet transfer mode stays the same. The droplet, on the other hand, diminishes as the droplet transfer frequency increases.

3.7 HYBRID GMAW

The combination of both laser and arc welding processes is known as laser–arc-based hybrid welding. Laser-arc hybrid welding improves the overall quality of the weld instead of considering either of the sources. This hybrid welding is an advanced welding technique; thus, it is widely used in the automobile, shipbuilding, pipeline, and telescopic industries [29]. Although the quality of the weld produced by the laser–arc hybrid is greater, there are more controllable process parameters, such as welding speed, laser power, wire feed speed, current, voltage, laser distance, laser angle, laser pulse rate, laser spot diameter, etc., making process control (such as predicting the shape of the weld) more complex. As will be discussed in a later section of the paper, these factors have a direct impact on the melting behavior of the electrode and base plate, causing the weld bead form to take on an unpredictable multi-curved profile and asymmetry. In car applications, which are a prospective scenario for this investigation, laser–arc hybrid welds are typically fillet-lap joints, which increase the complexity of the weld form. The shape and size of the weld beads are significant characteristics that affect strength, quality, and aesthetics. This is especially true for sheet welds used in automotive components, as exposed welds affect the component's overall look. Furthermore, to avoid puncturing a component, there is stringent control over the weld penetration of sheet welds in particular. The University of Kentucky created a novel welding technique called bypass coupling arc GMAW. In this endeavor, the University collaborated with Shandong University and Lanzhou University of Technology to further research on bypass coupling arc GMAW [26].

Gas metal arc welding with bypass coupling arc and laser hybrid welding are two effective forms of hybrid welding for thin plates. The British researcher W. Steen created the novel welding technique known as laser hybrid welding by creatively fusing two heat sources with various physical attributes and energy transmission mechanisms [26]. By fully using their respective advantages, the two distinct heat sources function concurrently at the same location and mutually compensate to generate a new, highly efficient heat source. Weld penetration is increased and the sensitivity to butt weld joint gaps is reduced via laser hybrid welding. The speed and deposition rate of welding may be increased by using two heat sources at once, but this approach is not widely adopted due to the laser device's large size, expensive cost, and high energy consumption [30, 31].

Figure 3.14 represents a schematic diagram of hybrid GMAW welding that includes the process of detecting spatial position and the process of using a high-speed camera for spatial positioning in arc-sourced MIG welding process. The diagram neatly explores the procedure with respect to the weld direction for efficiency in performing the hybrid welding.

Figure 3.15 illustrates the schematic diagram that represents the welding coalescence of arc welding and hybrid welding. First, as shown in Figure 3.15(a) and (b), the base material vaporizes, excites, and ionizes as a result of the high-temperature zone that the laser's keyhole effect creates in the arc cathode. In the arc plasma plume, the basic material transforms into microparticles. Second, the field emission is mostly responsible for the electron emission from the arc cathode due to the physical characteristics of Al alloys [32]. The cathode voltage drop increases as a result of the applied electromagnetic field produced by the laser beam on the keyhole's surface. The number of excited particles in the arc increases with increasing electron energy because it is easier for electrons to cross the potential barrier and join the arc. Thirdly, how the laser affects the arc impact ionization.

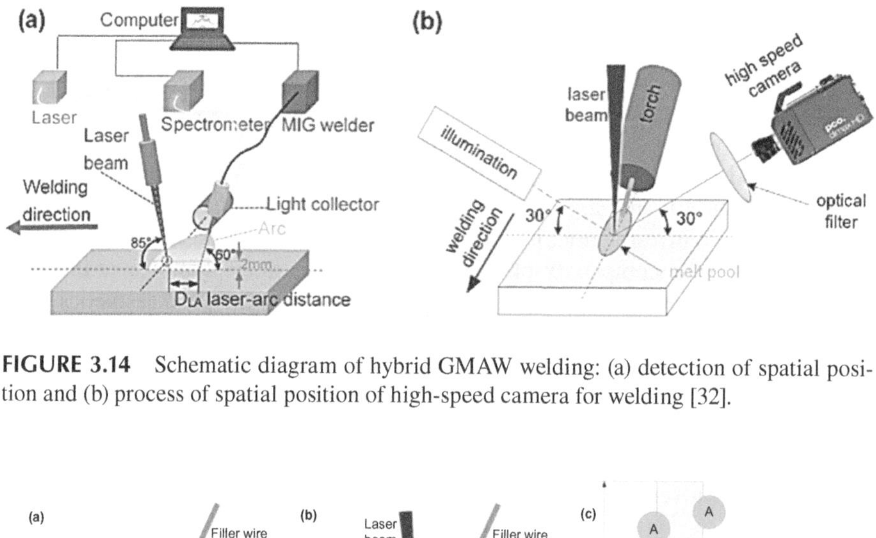

FIGURE 3.14 Schematic diagram of hybrid GMAW welding: (a) detection of spatial position and (b) process of spatial position of high-speed camera for welding [32].

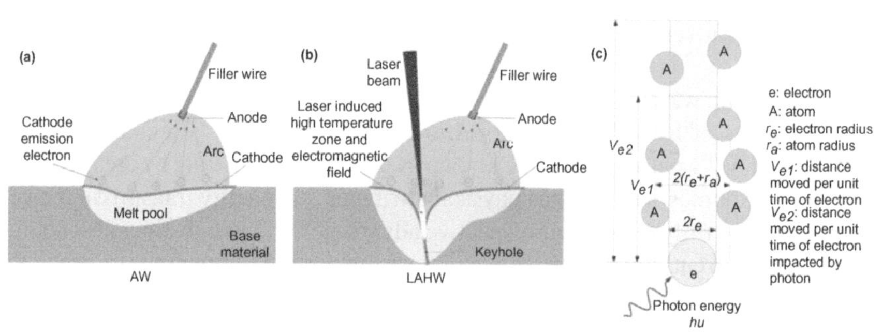

FIGURE 3.15 Schematic comparison of arc and laser sources: (a) arc welding and (b) laser–arc interaction [32].

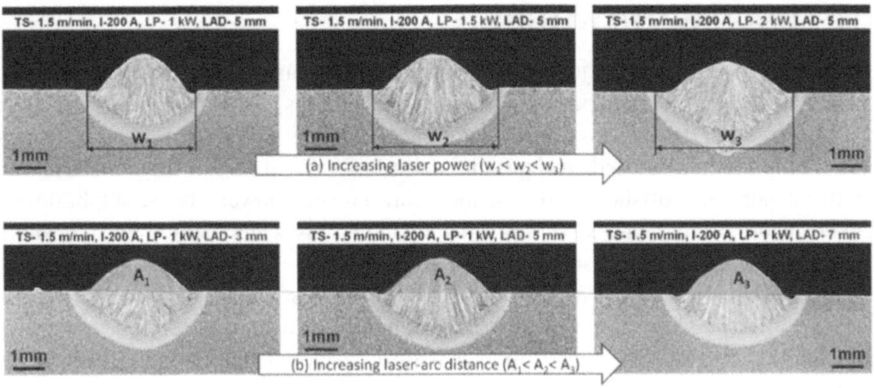

FIGURE 3.16 Effect of operating parameters on bead shape: (a) effect of laser power and (b) laser–arc distance [34].

It is interesting to note that the laser–arc distance and the laser beam have a mutual effect on deposition. Figure 3.16(b) shows that the deposition area and width increase with the laser–arc distance, defying the popular acceptance that the deposition would increase at a short distance. The paradox is that the laser placed close to the arc lowers the arc current. Laser energy is absorbed on the wire due to the laser's radiation [33]. Both the arc current and its contribution to the melting of the wire decrease at a fixed wire feed rate. As shown in Figure 3.16(b), an increase in the laser–arc distance causes an increase in the electrode melting rate, which in turn causes an increase in the deposition area [34]. The molten metal flow is affected by the laser beam, which in turn affects bead formation. The metal droplet is guided backward once it hits the melt pool's bottom. The heat is transferred to the back part by this constant flow, which causes the melt pool to lengthen and the bead to enlarge [35]. The findings of this study demonstrate the bead-widening effect, as illustrated in Figure 3.16(a). Bead width rises with increasing laser power (1, 1.5, and 2 kW). It is crucial to remember that while laser power alters the breadth, the height varies by a maximum of 0.19 mm, explaining the bead depression as a function of distance.

Important characteristics that affect the strength and quality of sheet metal welds include the size and form of the weld bead, which consists of the inner fusion boundary and the outer weld surface. Because of the asymmetry and other regulating factors, it is difficult to estimate the bead form in laser–arc hybrid fillet-lap joints using lower-order nonlinear analytical mathematical functions. The problem is solved by creating an artificial neural network that considers the inputs of voltage, current, laser power, welding speed, and wire feed speed. While the laser power has little effect, the welding speed and wire feed speed significantly influence the form of the bead. Due to base plate preheating and welding arc stabilization, the laser, despite its lower power, produces better weld dilution [29].

3.8 UNDERWATER WELDING

Underwater welding is the joining process of two similar or different metals that are placed underwater. It is a specific welding process that is performed while the welder is submerged in the underwater environment, often at elevated barometric pressures [36]. This method is used in the manufacture of several engineering materials in both onshore and offshore settings, including nuclear power plants, shipbuilding, and offshore oil and gas extraction. This technology is widely employed in onshore and offshore structure constructions due to the high productivity during the assembly procedures [37, 38]. The approach that works best for the welding process is heavily influenced by the material structure. It will have an impact on the material's mechanical characteristics and microstructure development. Thus, it will have an impact on the welded joint's outcome.

One of the greatest techniques for underwater structural emergency maintenance is the underwater hyperbaric dry GMAW procedure. The quality of GMAW reduces as ambient pressure rises. The primary cause of the welding process's instability is the shrinkage of the welding arc under high pressure [39]. Experiments were carried out to determine the primary determinants of the welding arc contraction. The findings demonstrate the phenomenon of arc contraction, with the axial and radial dimensions of the hyperbaric GMAW arc decreasing with increasing ambient pressure. Notably, the arc length also significantly decreases, increasing the likelihood of short-circuit drop transfer and compromising the stability of the welding process.

The welding technique described in the previous sections is the common welding technology used widely in the steel construction industry. In addition, the underwater welding process can employ the welding technique already mentioned. Nevertheless, in contrast to the standard welding procedure, the underwater welding method uses an arc that burns in bubbles, much as submerged arc welding does in underwater combustion. The layer encircling the welding electrode creates a sleeve to support the bubble while it burns. Applying a certain coating thickness and impregnating the electrode core with paraffin or another waterproof substance are required for the welding electrode to burn steadily underwater. The components of bubbles include hydrogen, oxygen, water vapor, and various oxides from turbid smoke, as well as bubbles produced by the burning of the electrode. It takes a higher arcing voltage and a current that is 15–20% more than the welding current in a conventional welding procedure to overcome the challenges of arc stability brought on by the lower temperature and water pressure.

Three techniques are typically available for underwater welding: local dry welding, wet welding, and dry welding. When doing wet welding, divers are often submerged in water to complete the weld directly, as shown in Figure 3.17(a) and (b), with the assistance of specialized equipment, the conditions of the underwater welding process are shown. The primary goal of the particular gear is to ensure the safety of the divers. Typically, it uses a similarly designed welding rod and a comparable procedure to regular welding. For example, the SMAW is typically used for submerged pipeline maintenance and underwater welding in the building of offshore

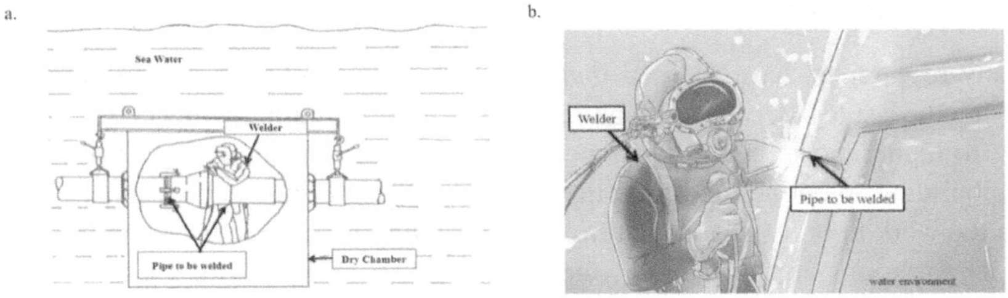

FIGURE 3.17 Underwater welding process: (a) dry welding and (b) wet welding [40].

platforms. Tungsten inert gas (TIG) and flux core arc welding (FCAW) are also often used for underwater welding applications.

Dry welding is also known as hyperbaric welding and is the second technique for underwater welding. The method of sealing a chamber around the weld component that has to be welded is called dry welding. The water is then forced out of the chamber, and a cavity is formed by filling it with argon gas, a mixture of helium and oxygen. This enables the weld to be done in a dry atmosphere. It is convenient to carry out TIG, MIG, FCAW, and SMAW in the specially constructed hyperbaric chamber. High-pressure dry welding and normal-pressure dry welding are the two common dry welding procedures. When using high-pressure dry welding, the inner pressure of the chamber is greater than the pressure in the welding area. In addition, when the water depth in the welding location increases, the inner pressure of the chamber rises as well, allowing the water to be expelled. Typically, flux-shielded arc welding and manual metal arc welding are used. The interior pressure of the chamber is kept at atmospheric pressure for normal-pressure dry welding. The depth of the water will not affect the welding process. To maintain the constant inner pressure (1 atm), the chamber layout for normal-pressure dry welding is more intricate.

A little cover and the welding flame are combined in the third underwater local dry welding technique [38, 40]. Water can be separated from the welding site by the welding shield gas. To safeguard the welding arc, the local cavity may move together with the welding torch during the procedure. Similar to wet welding, underwater local dry welding involves the diver performing the welding in the water. Usually, a specially made cover is used to keep water out of the weld zone. To isolate the welding arc from the surrounding water, gas is introduced into the cover. The rather dry climate is suitable for the welding procedure. There are typically three different types of dry welding techniques: large-scale cover dry welding, medium-scale cover dry welding, and small-scale cover dry welding. The water was ejected during the dry welding process using a large-scale cover, which makes it convenient to weld in a sizable hollow. A specially built cover is used in the medium-scale dry welding process, and the weld wire is easily introduced from a pre-made channel. Once the water has been forced out by the internal rotating airflow, the welding process may begin. The gas produced by the welding will be released simultaneously with the operation.

3.9 SUMMARY

Many researchers have focused their attention on the multiple-wire SAW process during the last few decades, and more recently, on the multiple-wire GMAW method, due to high deposition at higher speeds. These procedures are very helpful for the shipbuilding, pipeline, and pressure vessel manufacturing sectors, among others. On the work floor, twin-wire multiple-wire welding is an effective technique. Many of its variations are now in use, including tandem, TIME-twin, double-electrode, and indirect arc. Triple-wire, four-wire, and six-wire variants are still in the process of being completely implemented. The underwater environment makes the welding more complicated than the open-air welding process. The poor quality of the underwater welded joint is highly dependent on the content of hydrogen produced during underwater welding. Functional gradient materials and metallic components produced by additive manufacturing have also been produced in recent years using multiple-wire welding. The main issue with multiple-wire welding is arc instability, which is caused by the electromagnetic interaction of the arcs and is something that may be minimized but not completely removed. Control of arc stability and consequently its effects on energy distribution, deposition, melting, weld bead shape, microstructure, properties, etc., have been evaluated with the aid of sophisticated instrumentation and mathematical tools; however, numerous research efforts are necessary to fully comprehend the thermal, mechanical, and metallurgical aspects of multiple-wire welding.

REFERENCES

1. S. Goka, S. Q. Moinuddin, M. Cheepu, and A. K. Dewangan, "Welding practices in industry 5.0: Opportunities, challenges, and applications," in *Automation in Welding Industry*. Wiley, 2024, pp. 263–279. https://doi.org/10.1002/9781394172948.ch15.
2. T. Ashton, "Twin wire submerged arc welding," *Weld. J.*, vol. 33, pp. 350–355, 1954.
3. S. L. Mandelberg and V. E. Lopata. "High-speed twin-electrode submerged welding process." *Weld. Prod.*, vol. 2, pp. 25–28, 1966.
4. G. D. Uttrachi. "Multiple electrode system of SAW." *Weld. J.*, vol. 57, pp. 15–22, 1966.
5. E. Lassaline, B. Zajaczkowski, and T. H. North. "Narrow groove twin-wire GMAW of high-strength steel." *Weld. J.*, vol. 68, pp. 53–58, 1989.
6. T. Ueyama, T. Ohnawa, M. Tanaka, and K. Nakata, "Effect of torch configuration on high speed welding bead formation in tandem pulsed GMA welding process," *Q. J. JAPAN Weld. Soc.*, vol. 23, no. 3, pp. 383–391, 2005. https://doi.org/10.2207/qjjws.23.383.
7. R. Pablo Reis, D. Souza, and D. F. Filho, "Arc interruptions in tandem pulsed gas metal arc welding," *J. Manuf. Sci. Eng.*, vol. 137, no. 1, Feb. 2015. https://doi.org/10.1115/1.4028681.
8. S. Goka, M. Mallaiah, and M. J. Davidson, "Optimization of bead morphology for cold metal transfer wire arc additive manufacturing of AZ31 magnesium alloy wires," 2024, pp. 129–137. https://doi.org/10.1007/978-3-031-50240-8_27.
9. S. Goka, M. Manjaiah, and M. J. Davidson, "Study on mechanical, microstructural and corrosion analysis of wire arc additive manufactured AZ31 magnesium alloy," *Met. Mater. Int.*, 2024. https://doi.org/10.1007/s12540-024-01864-w.

10. G. Singh, A. K. Dewangan, M. F. Khan, and S. Q. Moinuddin. "Fundamental review on gas tungsten arc welding of magnesium alloys: challenges, innovations, and future perspectives." *Weld. World*, vol. 69, no. 9, pp. 2767–2787, 2025. https://doi.org/10.1007/s40194-025-02047-w.
11. S. Goka, S. N. Srirama, G. Divya Jyothi, S. Q. Moinuddin, and H. S. Shaik, "Welding-based 3D, 4D, 5D printing," in *Automation in Welding Industry*. Wiley, 2024, pp. 157–181. https://doi.org/10.1002/9781394172948.ch10.
12. J. D. Kulkarni, S. B. Goka, P. K. Parchuri, H. Yamamoto, K. Ito, and S. Simhambhatla, "Microstructure evolution along build direction for thin-wall components fabricated with wire-direct energy deposition," *Rapid Prototyp. J.*, vol. 27, no. 7, pp. 1289–1301, 2021. https://doi.org/10.1108/RPJ-04-2020-0085.
13. J. Tušek and M. Suban, "High-productivity multiple-wire submerged-arc welding and cladding with metal-powder addition," *J. Mater. Process. Technol.*, vol. 133, no. 1–2, pp. 207–213, Feb. 2003. https://doi.org/10.1016/S0924-0136(02)00235-2.
14. S. Q. Moinuddin and A. Sharma, "Multiple-wire welding GMAW and SAW," in *Advances in Welding Technologies for Process Development*, 1st edition. Boca Raton, FL : CRC Press/Taylor & Francis Group, 2019, pp. 1–22. https://doi.org/10.1201/9781351234825-1.
15. D. H. Kam, T. H. Lee, D. Y. Kim, J. Kim, and M. Kang, "Weld quality improvement and porosity reduction mechanism of zinc coated steel using tandem gas metal arc welding (GMAW)," *J. Mater. Process. Technol.*, vol. 294, no. June 2020, p. 117127, 2021. https://doi.org/10.1016/j.jmatprotec.2021.117127.
16. S. M. Adinarayanappa and S. Simhambhatla, "Twin-wire welding based additive manufacturing (TWAM): Manufacture of functionally gradient objects," *Rapid Prototyp. J.*, vol. 23, no. 5, pp. 858–868, Aug. 2017. https://doi.org/10.1108/RPJ-09-2015-0126.
17. S. M. Adinarayanappa and S. Simhambhatla, "Determination of process parameter for twin-wire weld-deposition based additive manufacturing," Volume 4: 19th Design for Manufacturing and the Life Cycle Conference; 8th International Conference on Micro- and Nanosystems, American Society of Mechanical Engineers, Aug. 2014. https://doi.org/10.1115/DETC2014-34658.
18. M. Kumar, S. Quadir Moinuddin, S. S. Kumar, and A. Sharma, "Discrete wavelet analysis of mutually interfering co-existing welding signals in twin-wire robotic welding," *J. Manuf. Process.*, vol. 63, pp. 139–151, Mar. 2021. https://doi.org/10.1016/j.jmapro.2020.04.048.
19. S. Q. Moinuddin and A. Sharma, "Arc stability and its impact on weld properties and microstructure in anti-phase synchronised synergic-pulsed twin-wire gas metal arc welding," *Mater. Des.*, vol. 67, pp. 293–302, Feb. 2015. https://doi.org/10.1016/j.matdes.2014.11.052.
20. M. Schnick, G. Wilhelm, M. Lohse, U. Füssel, and A. B. Murphy, "Three-dimensional modelling of arc behaviour and gas shield quality in tandem gas–metal arc welding using anti-phase pulse synchronization," *J. Phys. D. Appl. Phys.*, vol. 44, no. 18, p. 185205, May 2011. https://doi.org/10.1088/0022-3727/44/18/185205.
21. A. Hamrani, F. Z. Bouarab, A. Agarwal, K. Ju, and H. Akbarzadeh, "Advancements and applications of multiple wire processes in additive manufacturing: A comprehensive systematic review," *Virtual Phys. Prototyp.*, vol. 18, no. 1, pp. 1–34, 2023. https://doi.org/10.1080/17452759.2023.2273303.
22. K. Treutler and V. Wesling, "The current state of research of wire arc additive manufacturing (Waam): A review," *Appl. Sci.*, vol. 11, no. 18, 2021. https://doi.org/10.3390/app11188619.

23. H. Hamed Zargari, K. Ito, M. Kumar, and A. Sharma, "Visualizing the vibration effect on the tandem-pulsed gas metal arc welding in the presence of surface tension active elements," *Int. J. Heat Mass Transf.*, vol. 161, p. 120310, 2020. https://doi.org/10.1016/j.ijheatmasstransfer.2020.120310.
24. J. Zhang, S. Chen, H. Zhao, Y. Yu, and M. Liu, "Designing a multi-output power supply for multi-electrode arc welding," *Electronics*, vol. 12, no. 7, p. 1702, Apr. 2023. https://doi.org/10.3390/electronics12071702.
25. Z. Ma, M. Zhuang, and M. Li, "Effect of main arc voltage on arc behavior and droplet transfer in tri-arc twin wire welding," *J. Mater. Res. Technol.*, vol. 9, no. 3, pp. 4876–4883, May 2020. https://doi.org/10.1016/j.jmrt.2020.03.007.
26. Y. Chen, C. Fang, Z. Yang, J. Wang, G. Xu, and X. Gu, "Cable-type welding wire arc welding," *Int. J. Adv. Manuf. Technol.*, vol. 94, no. 1–4, pp. 835–844, 2018. https://doi.org/10.1007/s00170-017-0943-4.
27. C. Fang, Y. Chen, Z. Yang, J. Wang, M. Wu, and K. Qi, "Cable-type welding wire submerged arc surfacing," *J. Mater. Process. Technol.*, vol. 249, pp. 25–31, Feb. 2017. https://doi.org/10.1016/j.jmatprotec.2017.05.020.
28. Z. Yang, C. Fang, M. Wu, K. Qi, Y. Chen, and R. Ma, "The mechanisms of arc coupling and rotation in cable-type welding wire CO2 welding," *J. Mater. Process. Technol.*, vol. 255, pp. 443–450, Dec. 2018. https://doi.org/10.1016/j.jmatprotec.2017.12.032.
29. P. Kochar, A. Sharma, T. Suga, and M. Tanaka, "Prediction and control of asymmetric bead shape in laser-arc hybrid fillet-lap joints in sheet metal welds," *Lasers Manuf. Mater. Process.*, vol. 6, no. 1, pp. 67–84, 2019. https://doi.org/10.1007/s40516-019-0081-y.
30. S. Ghosal and S. Chaki, "Estimation and optimization of depth of penetration in hybrid CO2 LASER-MIG welding using ANN-optimization hybrid model," *Int. J. Adv. Manuf. Technol.*, vol. 47, no. 9–12, pp. 1149–1157, Apr. 2010. https://doi.org/10.1007/s00170-009-2234-1.
31. X. Zhang, Z. Zhao, C. Wang, F. Yan, and X. Hu, "The effect of external longitudinal magnetic field on laser-MIG hybrid welding," *Int. J. Adv. Manuf. Technol.*, vol. 85, no. 5–8, pp. 1735–1743, Jul. 2016. https://doi.org/10.1007/s00170-015-8035-9.
32. C. Zhang, M. Gao, and X. Zeng, "Influences of synergy effect between laser and arc on laser-arc hybrid welding of aluminum alloys," *Opt. Laser Technol.*, vol. 120, p. 105766, Dec. 2019. https://doi.org/10.1016/j.optlastec.2019.105766.
33. J. Wang, H. Nishimura, K. Fujii, S. Katayama, and M. Mizutani, "Study of improvement of gap tolerance in laser MIG arc hybrid welding of aluminium alloy," *Weld. Int.*, vol. 23, no. 10, pp. 723–733, Oct. 2009. https://doi.org/10.1080/09507110902836879.
34. A. Kapil, T. Suga, M. Tanaka, and A. Sharma, "Towards hybrid laser-arc based directed energy deposition: Understanding bead formation through mathematical modeling for additive manufacturing," *J. Manuf. Process.*, vol. 76, pp. 457–474, Apr. 2022. https://doi.org/10.1016/j.jmapro.2022.02.027.
35. Y. Naito, M. Mizutani, and S. Katayama, "Observation of keyhole behavior and melt flows during laser-arc hybrid welding," International Congress on Applications of Lasers & Electro-Optics, Laser Institute of America, 2003, p. 1005. https://doi.org/10.2351/1.5059966.
36. J. H. Nixon, C. J. Allum, and J. M. Lowes, "Underwater welding - a review," pp. 147–172, Jan. 1982. https://doi.org/10.2478/v10077-008-0040-3.
37. G. Fu, "Underwater welding," in *Encyclopedia of Ocean Engineering*. Singapore: Springer Nature Singapore, 2022, pp. 2087–2093. https://doi.org/10.1007/978-981-10-6946-8_227.

38. S. Q. Moinuddin *et al.*, "Research progress in underwater welding: Techniques, materials, advancements, and challenges," *Weld. World*, May 2025. https://doi.org/10.1007/s40194-025-02068-5.
39. J. Huang, W. Guo, J. Huang, L. Xue, J. Wang, and Y. Zou, "Arc contraction behavior of GMA welding process based on change of arc length at hyperbaric air condition," 2020, pp. 47–60. https://doi.org/10.1007/978-981-15-6922-7_3.
40. E. Surojo, E. D. W. S. Putri, E. P. Budiana, and Triyono, "Recent developments on underwater welding of metallic material," *Procedia Struct. Integr.*, vol. 27, no. 2019, pp. 14–21, 2020. https://doi.org/10.1016/j.prostr.2020.07.003.

4 Process Monitoring Techniques

4.1 BACKGROUND

Welding is a complex and dynamic process that inherently involves uncertainties, making real-time monitoring and control essential to achieve consistent and high-quality results. With recent advancements in power source technologies, modern welding systems have become more sophisticated, involving multiple controllable input parameters that must be regulated to minimize disturbances and ensure optimal output quality.

The welding process is often broken down into smaller, manageable subprocesses to manage this complexity, as illustrated in Figure 4.1. Each subprocess can be monitored individually for specific inputs, outputs, and potential disturbances. Key parameters typically tracked include welding current and voltage, arc characteristics, temperature, metal transfer mode, weld pool geometry, weld seam integrity, penetration depth, wire feed rate, shielding gas flow, and the presence of defects.

Welding process monitoring can generally be categorized into three stages:

- Pre-process monitoring: Evaluates setup parameters like material properties, joint configuration, and equipment readiness.
- In-process monitoring: Tracks live data such as arc stability, current/voltage behavior, pool dynamics, and metal transfer in real time.
- Post-process monitoring: Assesses the quality of the weld through visual, thermal, and NDT (non-destructive testing) methods.

In the context of Additive Manufacturing (AM), especially processes such as wire arc additive manufacturing (WAAM), monitoring can be structured at three scales:

- Melt pool level—Observes temperature gradients, melt pool dimensions, and heat-affected zone (HAZ) behavior to determine local quality and process stability.
- Single-layer level—Analyzes layer geometry, thickness, surface pattern, thermal distribution, and defect presence to ensure inter-layer bonding and dimensional accuracy.
- Build level—Tracks global shape evolution, component distortion, and thermal accumulation across the entire build to manage overall structural integrity.

DOI: 10.1201/9781003778844-4

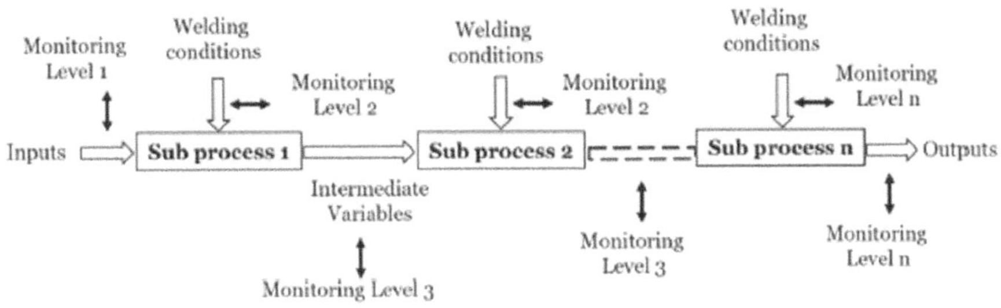

FIGURE 4.1 Effective control and monitoring of welding processes [1].

Monitoring relies heavily on a range of contact and non-contact sensors, which convert physical process phenomena into measurable signals. These include thermocouples, infrared cameras, acoustic sensors, laser scanners, and visual imaging systems. A diverse sensor suite is essential to monitor and control both welding and AM processes, ensuring that the resulting components meet mechanical, metallurgical, and dimensional specifications.

Despite technological advancements, many systems still depend heavily on manual sensor setup and calibration, which can introduce errors and prolong setup times. As manufacturing environments grow increasingly integrated with multiple processes operating concurrently, automated and intelligent monitoring systems are becoming critical for quality assurance and process optimization.

The total welding process chain is undergoing rapid digital transformation, driven by the integration of advanced technologies such as artificial intelligence (AI), machine learning (ML), deep learning, and big data analytics [2]. This digital evolution is commonly referred to as the concept of the "digital twin," which aims to create a virtual replica of the physical welding system for real-time simulation, prediction, and control.

This transformation is closely aligned with the principles of Industry 4.0 and the emerging vision of Industry 5.0, both of which promote smart, interconnected manufacturing systems that prioritize autonomy, adaptability, and human-centric design [3, 4]. In these frameworks, automation powered by the Internet of Things (IoT) and machine-to-machine (M2M) communication enables real-time data exchange, decentralized decision-making, and enhanced self-organization of the welding and additive manufacturing (AM) processes.

As illustrated in Figure 4.2, the integration of multi-sensor networks into the digitized workflow allows for comprehensive process monitoring. This enhances control over thermal cycles and material properties, resulting in improved property homogeneity and dimensional consistency of the fabricated components.

This chapter presents various process monitoring techniques, focusing on the practical application of modern sensors in welding and AM. It emphasizes how these sensing technologies support real-time data acquisition and decision-making for optimized process stability, quality control, and defect prevention—core goals in today's intelligent manufacturing environments.

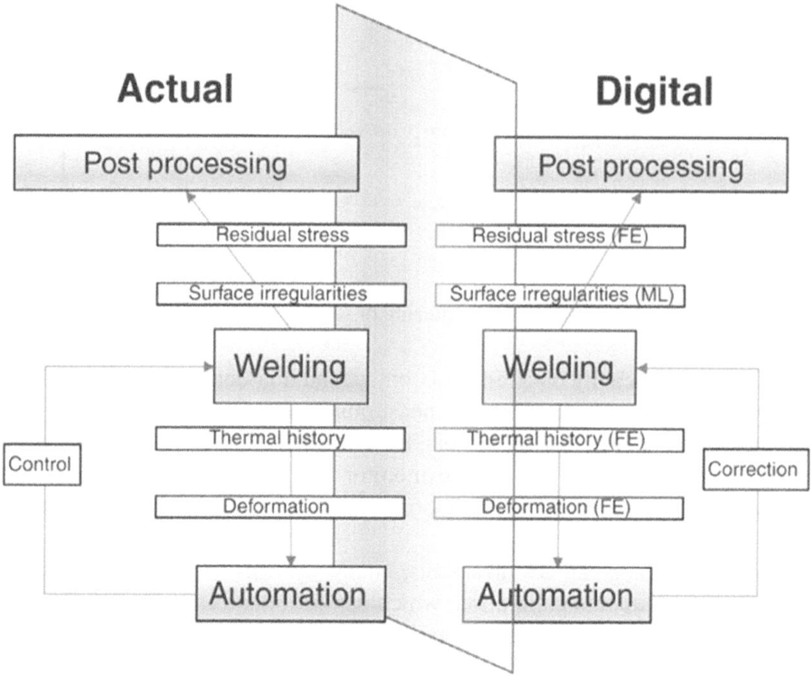

FIGURE 4.2 Concept map of welding digital twin [5].

4.2 CURRENT AND VOLTAGE MONITORING

Welding current and voltage are critical parameters that govern heat input, weld bead geometry, and microstructural evolution in arc welding processes. In consumable electrode welding, the welding current is directly linked to the wire feed speed, while the welding voltage corresponds to the arc length.

Depending on the application, arc welding utilizes either constant current (CC) or constant voltage (CV) power sources. Both direct current (DC) and alternating current (AC) are employed in industry, depending on the material type and desired arc characteristics. However, due to fluctuations in arc resistance, both current and voltage tend to vary during operation, especially in consumable electrode processes where molten metal droplets form and detach at the electrode tip [5].

4.2.1 Current Sensing Techniques

Current can be monitored using Hall-effect sensors, shunt resistors, or clamp meters:

Clamp meters (Figure 4.3) are widely used due to their non-intrusive design. In the case of AC measurement, the device works on the principle of a current transformer, where the conductor acts as the primary winding. The secondary current (I_s) is induced by the magnetic flux generated by the primary current (I_p) and follows the relation shown below:

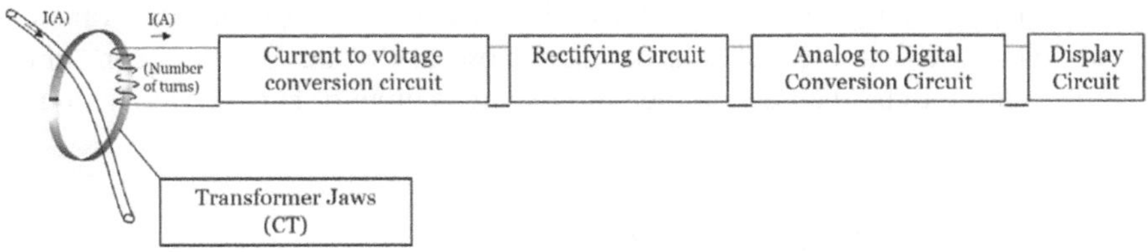

FIGURE 4.3 Clamp meter for measuring AC [5].

$$I_s = I_p/N \ldots\ldots\ldots\ldots \quad (4.1)$$

where N is the number of turns in the secondary winding.

Shunt resistors function based on Ohm's law. These are reliable and cost-effective options for both AC and DC measurements. The voltage drop across a resistive element (typically Manganin, chosen for its stable resistance and high conductivity) is measured to determine the current.

4.2.2 Voltage Sensing Techniques

Voltage measurement in welding is typically done using resistive or capacitive sensors. The resistive voltage sensors utilize either a voltage divider or bridge circuit configuration. Actually, in a voltage divider, two resistors scale down the high welding voltage to a measurable level. One resistor acts as the reference, and the other as the sensing element.

In a bridge circuit (Figure 4.4a), four resistors are arranged in a Wheatstone configuration, allowing precise detection of voltage differences due to resistance variation.

Capacitive voltage sensors (Figure 4.4b) consist of two conductors separated by an insulating dielectric. When voltage is applied, the system behaves like a capacitor, and the change in capacitance correlates with the applied voltage. This method

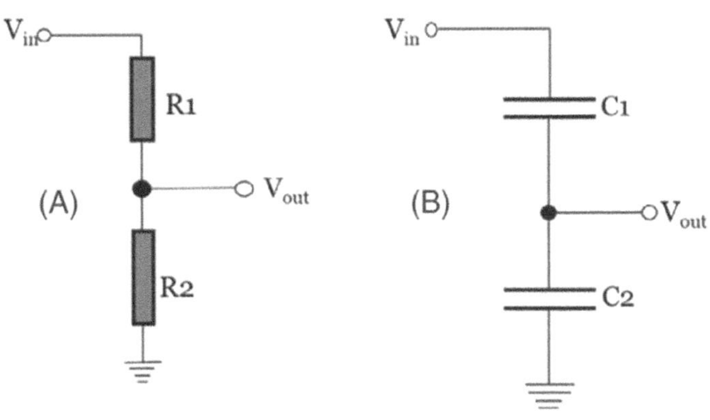

FIGURE 4.4 Types of voltage sensors: (a) resistive type and (b) capacitor type [5].

is especially suitable for high-voltage, high-frequency applications. Although ideal voltage measurement should be taken close to the arc, practical constraints often require it to be recorded at the power source output or between the torch and workpiece.

4.3 WIRE FEED RATE MONITORING

The wire feed rate is a critical parameter in gas metal arc welding (GMAW) as it directly influences the welding current, metal deposition rate, and arc stability. Traditional GMAW setups utilize a continuous wire feed mechanism; however, advancements in monitoring and control have introduced alternative methods like pulsed wire feeding for enhanced process stability and performance. To explore these alternatives, a custom-built apparatus was developed, superimposing a linear pulsating motion over the conventional continuous wire feed. This modification enables the wire to move in a controlled in-and-out pattern rather than at a constant speed. The goal is to achieve short-circuit transfer even at higher arc voltages, which are typically unsuitable for short-circuiting transfer modes using conventional methods [6]. The methodology was validated experimentally under both constant current (CC) and constant voltage (CV) power supply modes. Various pulsation frequencies were tested and compared to standard continuous wire feed conditions. This comparison helped in evaluating the influence of pulsation on arc behavior, metal transfer characteristics, and weld quality.

As illustrated in Figure 4.5, the pulsation device enables synchronized actuation between the wire feeding actuator and the wire tip, offering precise control during arcing. Notably, Figure 4.5 presents the coordination mechanism without directly referencing wire feed rate, emphasizing the mechanical aspect of wire pulsation and its timing during welding. This technique moves beyond traditional waveform control by physically altering the wire movement, providing a new approach to achieve desirable transfer modes (especially short-circuiting) in regimes where it is otherwise challenging.

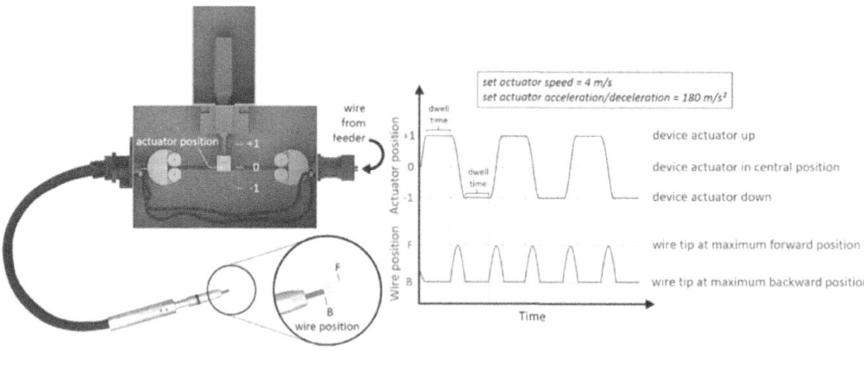

FIGURE 4.5 Pulsation device and schematic diagram of the synchronization between the device actuator and the wire tip during welding [2].

4.4 TEMPERATURE MONITORING

Thermal management, or regulating the temperature that determines the quality of the weld, is another crucial component of arc welding and AM processes [3]. Thermocouples were incorporated to monitor the inter-pass temperatures between the layer upon layer deposition in the GMAW process [4] to have control of microstructure and mechanical properties in the build direction. So, temperature is a key factor in the mechanical characteristics and solidification of the material. Several kinds of sensors fall roughly into two categories: contact type and non-contact type sensors. They can be used to monitor temperature.

Three further categories are used for further classification of contact-type sensors: electrical-resistive, thermo-electrical, and electro-mechanical. Thermocouples are the most common type of thermo-electrical equipment used to detect welding temperatures, which range from –200°C to 3000°C. Two junctions made of different metals make up a thermocouple; one serves as the measurement junction (hot) and the other as the reference (cold). As shown in Figure 4.6, a voltage drop is produced when the two junctions' respective temperatures differ. This voltage drop is utilized to determine the temperature.

Table 4.1 illustrates the many thermocouple material combinations that are available depending on the temperature range. Applications requiring the K-type thermocouple's exceptional sensitivity to tolerate high temperatures include welding and AM.

Convection and radiation are used by the non-contact type temperature sensors for monitoring temperature variations. Three types of radiation detectors are distinguished: (a) thermal detectors, (b) optical pyrometers, and (c) photon or quantum detectors. As shown in Figure 4.8(a), the temperature is determined using an optical pyrometer by comparing the luminosities of the emitting source and reference light. The radiation coming from the source is captured by the objective optical lens. A lens is employed to concentrate the heat radiation onto the reference light. The reference lamp's current and intensity fluctuate when the observer modifies the rheostat settings. When the brightness of the temperature source and the filament are equal,

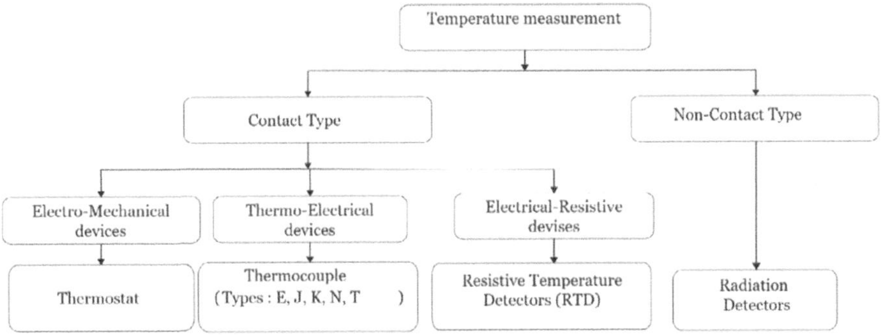

FIGURE 4.6 Flow chart of temperature measuring techniques [5].

TABLE 4.1
Standard Combinations of Thermocouples [5]

Letter Type	Material Composition	Temperature Range (°C)	Output (μV/°C)
B	Pt-30 Rh/Pt-6 Rh	100 to 1750	5 at 1000°C
E	Ni-Cr alloy/Cu-Ni alloy	−268 to 800	68 at 100°C, 81 at 500°C, 77 at 900°C
J	Fe/Cu-Ni alloy	−19 to 700	26 at 190°C, 63 at 800°C
K	Ni-Cr alloy	−250 to 1100	40 (250–1000°C), 35 at 1300°C
N	Ni-Cr-Si (Nicrosil)/Ni-Si-Mg (Nisil) alloy	0 to 1250	37 at 1000°C
R	Pt-13 Rh/Pt	0 to 1600	10 at 1000°C
S	Pt-10 Rh/Pt	0 to 1500	6 (0–100°C)
T	Cu/Cu-Ni alloy	−262 to 850	15 at 200°C, 60 at 350°C

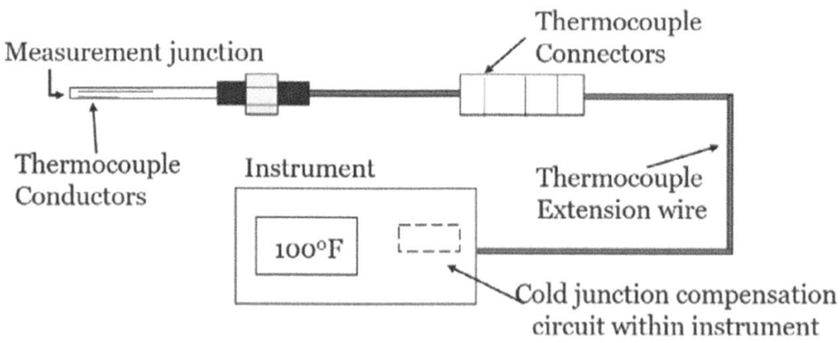

FIGURE 4.7 Construction of a thermocouple [5].

the current is calibrated on the temperature scale so that the filament cannot be visually differentiated. The temperature source is hotter than the filament, and vice versa if the filament is darker than the source's brightness.

A bolometer, thermopile, and pyroelectric detectors are a few examples of devices that may sense an increase in detector temperature caused by the thermal detector's conversion of absorbed electromagnetic radiation into heat energy. Pyroelectric detectors are among those that are frequently utilized in welding procedures. Bolometers and thermopiles react more slowly than pyroelectric detectors. Figure 4.8(b) illustrates the design and operation of a pyroelectric detector. Two electrodes filled with pyroelectric material such as lithium tantalite or triglycine sulfate make up the sensitive element. The radiation from the item is absorbed by the sensitive element, which modifies the pyroelectric material's surface electric charge. A preamplifier processes the electrical signal that the detector generates, which is proportional

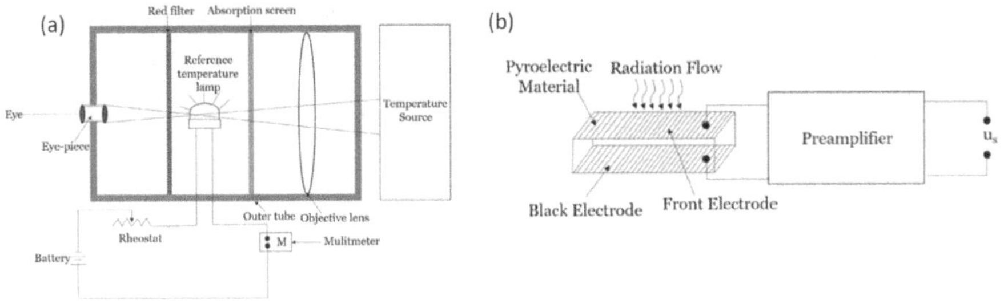

FIGURE 4.8 Construction of (a) an optical pyrometer and (b) a pyroelectric detector [5].

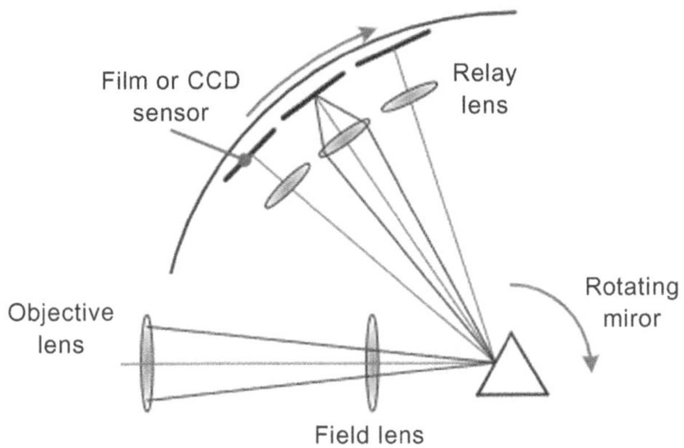

FIGURE 4.9 Working principle of high-speed camera: operation of the rotating mirror [6].

to the charge change rate. The photo effect is the basic principle for the operation of quantum detectors such as photoemissive, photoconductive, and photovoltaic, which measure the electron excitation to conduction states by incoming photons. However, photon detectors work by releasing or moving electrical charge carriers to record each photon's reaction. Because excited electrons go back to the ground state, quantum detectors are less efficient than thermal detectors, while having a higher sensitivity.

Temperatures can also be measured by recording videos. A high-speed camera that captures in slow motion is another crucial optical tool for studying transient phenomena. Figure 4.9 illustrates how high-speed cameras, in place of shutters, may capture pictures at up to 250,000 frames per second by passing the image via a revolving prism or mirror.

A slit must be placed on the input optical path and the relay lenses removed to operate a rotating mirror camera in streak mode. A maximum sweep speed of 490 ps/pixel has been documented with this device [7]. When measuring the rate of dimensional change in events of uniform growth, such as an expanding sphere, streak-mode imaging works well.

To prevent misunderstandings and offer the most information possible, streak and framing records are frequently used in tandem. For this reason, streak-mode cameras

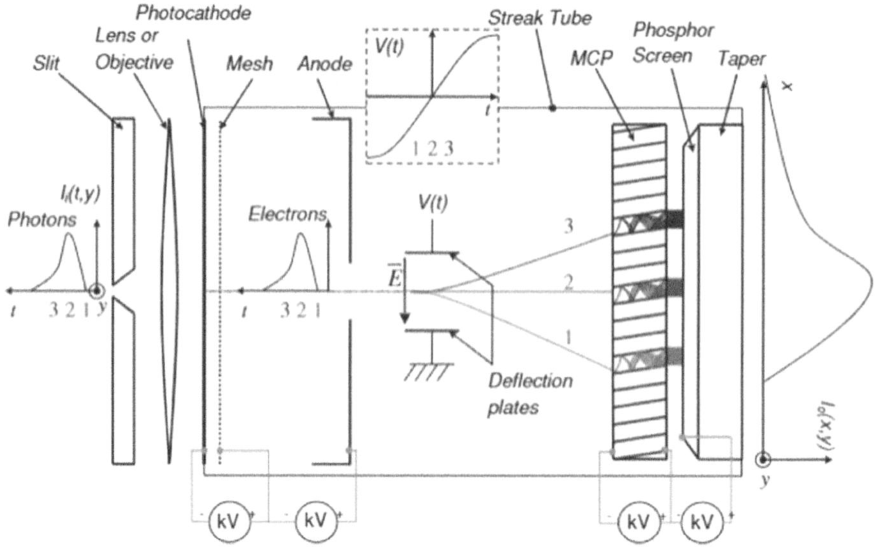

FIGURE 4.10 Working principle of a conventional streak camera [6].

and framing are complementary rather than similar [8]. The centrifugal force exerted on the mirror by the extremely high rotation speed limits the capabilities of spinning mirror technology. When the rotation speed surpasses 2 million rpm, the mirror will explode even if light metals like beryllium are used [9]. It is necessary to employ an entirely new technology built around a vacuum tube to get around this restriction. The picosecond-order temporal resolution may be attained with the standard streak camera, the fastest direct light measuring apparatus now in use. Figure 4.10 shows the basic workings of a traditional streak camera. A high-speed video camera may be configured to operate in streak mode by turning on multiple rows of the sensor array and concentrating the relevant spatial data on those rows. A demonstration of this technology is given by Parker et al. [10], where a sweep speed of 7.5 mm/μs and a temporal resolution of 1 μs are achieved, obviously restricted by the output data rate.

Conventional streak cameras and rotating mirror cameras have disadvantages in addition to their exceptional performance: they are expensive, large, brittle, and difficult to make. Furthermore, a temporal resolution of around 1 ns is enough in many applications. Solid-state technology can provide intriguing options for various applications. A normal streak camera typically has a spatial resolution of 25–100 μm.

4.5 WELD POOL MONITORING

Xia et al. (2020) monitored the real-time width of the molten pool, regulated the width of metal deposition in WAAM, and created a model-free adaptive iterative learning control system [11]. Usually non-contact, optical sensors can sense and regulate a range of welding characteristics, including weld pool, weld seam, and arc length. The conventional welding control system, shown in Figure 4.11, makes use of an optical sensor whose output is used to regulate the welding process. The optical sensor works by sensing light energy.

Electro-optic sensors, charge-coupled devices (CCD) sensors, complementary metal-oxide-semiconductor (CMOS) sensors, and high-speed cameras are among the sensors that are often used in welding operations. Light or electromagnetic energy is sensed by the electro-optic sensor, commonly known as a photodetector. Materials such as germanium, silicon, or alloys of indium and antimony, mercury, telluride, and cadmium are used to make photodetectors. These can function in either the photovoltaic or photoconductive modes. The conductivity of the detector material is altered by the photoconductive detectors' absorption of photons that raise electrons. As shown in Figure 4.12(a), the photoconductive detector is biased using the load resistor. In the absence of light, photovoltaic detectors are also referred to as photodiodes. The operation of a photovoltaic detector is shown in Figure 4.12(b). By absorbing photons, these detectors form an electron–hole pair across the p–n junction, which results in a photocurrent that charges the capacitor in proportion to the received radiation (without the need for an external bias).

Charge-coupled device (CCD) sensors are light-sensitive devices like photodiodes. They are constructed from silicon, which is frequently used in digital cameras. The near-infrared and visible light spectra are covered by the sensitivity range. The light intensity is detected by the CCD sensor and subsequently recorded as an output voltage signal. The inability of the CCD sensor to detect color light may be overcome

FIGURE 4.11 The optical sensor used in a control system of the welding process [5].

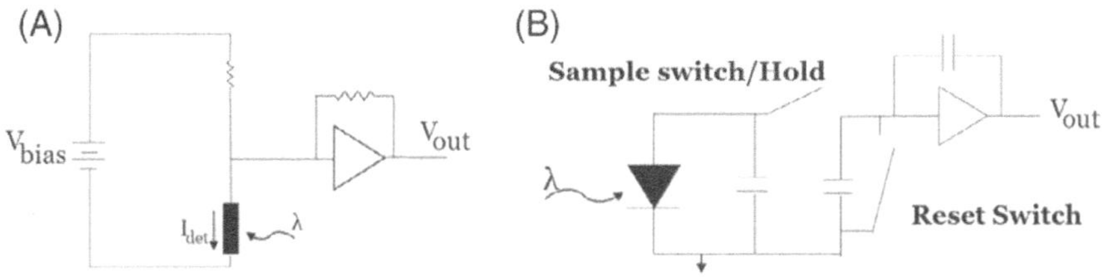

FIGURE 4.12 Optical sensors: (a) photo conductive detector and (b) photovoltaic detector [5].

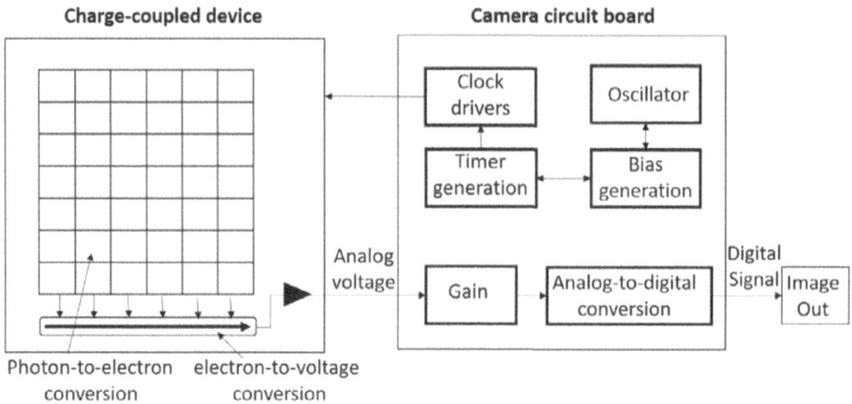

FIGURE 4.13 CCD sensor: working principle [5].

by applying color filters (red, green, or blue) to each pixel and then determining the information that is lacking. There are three fundamental methods for scanning images: area, line, and point scanning. Four alternative designs are used by the CCD area image sensor, based on transfer methods: frame transfer (FT), full-frame transfer (FFT), interline transfer (IT), and frame interline transfer (FIT). The types of detectors used in these transfer methods such as metal oxide semiconductors (MOS), photo capacitors, and photodiodes differ. The two most popular transfer methods for CCD sensors are FT and FFT. The idea behind all of the transfer techniques is the same: for every pixel in the picture array, the light signal is converted into an output voltage. Figure 4.13 illustrates the three steps that the CCD image sensor goes through to function. Exposure is the initial stage, in which light is transformed into an electric charge inside a pixel. The charge is transferred to the silicon substrate in the second process, known as charge transfer: converting the charge to the output voltage and amplifying it is the last stage.

4.6 WELD ACOUSTIC

Acoustic sensors like microphones are capable of converting sound pressure waves into electrical signals. The choice of microphone for a given monitored process depends on several characteristics, including dynamic range, size, sensitivity, and bandwidth. The pressure change created in the air is detected by the microphone. Humans can perceive sound waves in the frequency range of 20 Hz to 20 kHz. Ultrasonic waves are defined as sound waves with a frequency higher than 20 kHz. Only a liquid medium can effectively transmit the ultrasonic wave frequency. The sound pressure levels (SPL) in AM and welding processes can forecast several aspects of the operation, such as the kind of metal transfer, weld pool, weld flaws, and arc behavior. Decibels (dB) are used to express the SPL. There are several kinds of microphones, such as condenser microphones (capacitor), piezoelectric

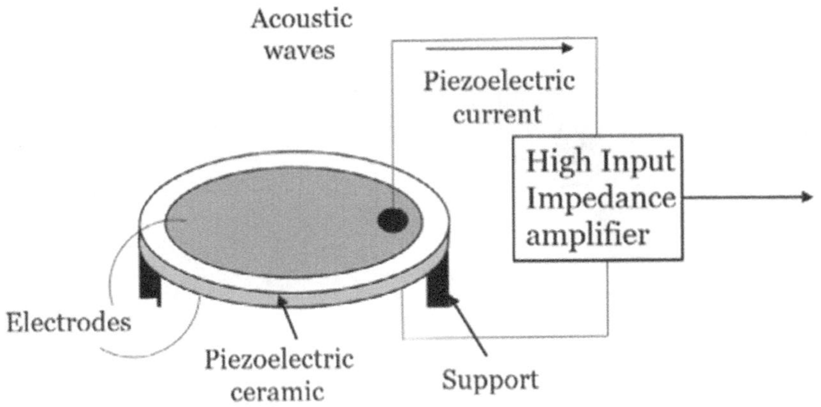

FIGURE 4.14 Microphone: piezoelectric type [5].

microphones, electret microphones, resistive microphones, and dynamic microphones (moving coil and ribbon varieties).

Figure 4.14 depicts a typical microphone. Piezoelectric microphones are the most commonly utilized kind of microphone among these others. The electrically conducting electrode of this microphone is made up of piezoelectric ceramic material on both ends. Mechanical forces are converted into electrical charges by the piezoelectric crystal. This sensor can also capture and process the emitted acoustic emission (AE).

4.7 BEAD GEOMETRY MONITORING

Researchers and manufacturers can obtain real-time insights into the layer-by-layer control of metal deposition in the directed energy deposition process by online monitoring of weld bead geometry. Optimization of process parameters for bead geometry is necessary to ensure the proper weld deposition phenomenon [12, 13]. A gray-based online monitoring system (GOM) was developed in the present study for monitoring the height and depth of weld beads [14]. This study established an optimization gray model (OGM) (1, N)-based online monitoring system to measure the height and depth of metal deposition in each layer [15]. Fewer information training samples are required for the OGM (1, N). AM was used to create online and offline tracking and modeling algorithms for metal-deposited layer geometries and their quality. A closed-loop control system integrating a PID controller and a passive vision camera was created by Xiong et al. [16]. Based on the height of the previous and current layers detected by the vision camera, the PID controller decreased height errors by adjusting the wire feed speed. According to Farias et al., the wall geometry in AM is impacted by the inter-pass temperature [17]. They used artificial neural networks (ANN) with FEM simulation. Xia et al. anticipated the height and surface roughness in WAAM using an enhanced adaptive neuro-fuzzy inference method [18]. Figure 4.15 represents the CMT-based WAAM process and monitoring system for the layer

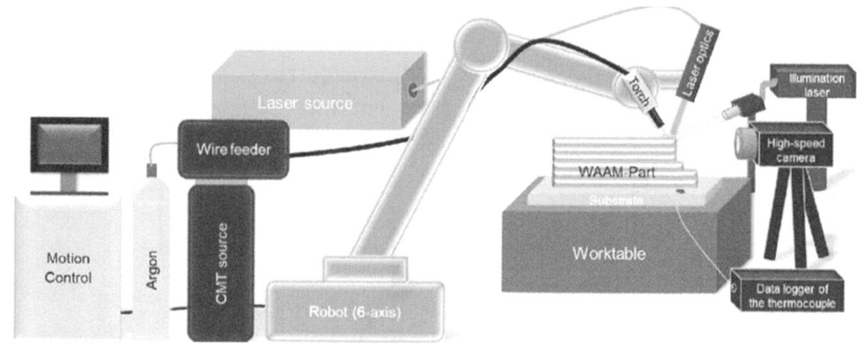

FIGURE 4.15 CMT-based WAAM process: weld bead monitoring setup [19].

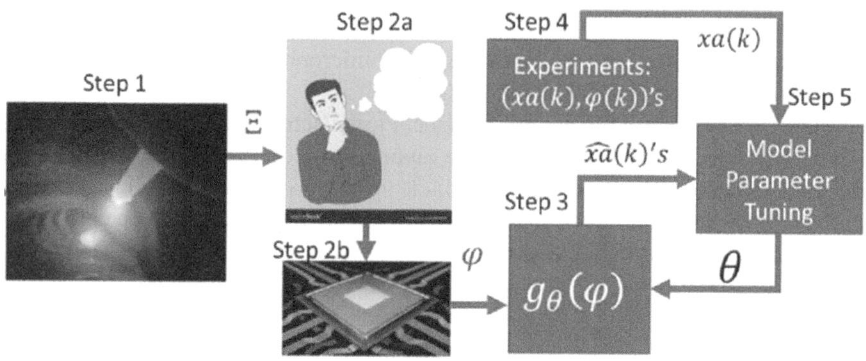

FIGURE 4.16 Key elements to develop models to monitor weld penetration of bead geometry [21].

upon layer deposition, and the measuring and monitoring process of surface roughness of the deposited weld beads. Thus, as a case, bead geometry and surface profile can also be monitored in the GMAW process.

Several methods are being researched to accomplish precise automation and control of the GMAW process. One major issue with reliable control is the online measurement of the deposited bead geometry. To detect the height and breadth of beads in real time by depositing thin walls [20], a passive vision-sensing system with two cameras and composite filtering algorithms was presented. Throughout the multilayer deposition process, the nozzle-to-tip-surface distance was tracked to remove any accumulated height inaccuracies. To extract feature parameters, a variety of image processing methods were used and described. For the monitoring system, a calibration process was provided. The online measuring method for bead geometry in layered additive manufacturing has been confirmed to be effective based on validation studies. Figure 4.16 illustrates the key elements to develop models to monitor weld penetration.

It is essential to measure the transient temperature or to keep an eye on temperature variations while welding to ensure that sound welds are produced and the process is of high quality. To determine the heat source model parameters based on the welding circumstances and to acquire the thermal cycle, thermocouples are employed to detect the temperature change in the workpiece's body [22]. For a more accurate description of the process, the heat source model parameters for various welding methods, such as single-wire welding, twin-wire welding [23], and square waveform welding [24], must be modified. The relationship between welding settings and weld qualities, such as bead form, may be precisely explained by the heat source models.

4.8 SUMMARY

With the rapid progression of Industry 4.0 and sensor technologies, welding is transitioning from a craft to a data-driven science. This transformation requires a multi-sensor-enabled welding environment capable of extracting, analyzing, and integrating real-time information from the process to ensure quality, efficiency, and repeatability.

A wide variety of sensors are currently utilized in arc welding monitoring systems. Current and voltage sensors are essential for determining welding power, analyzing metal deposition efficiency, maintaining arc stability, and identifying weld defects. Temperature sensors, such as thermocouples and infrared detectors, monitor arc and weld pool temperatures, providing insights into the thermal cycle, which is crucial for microstructure and phase transformation control.

Optical sensors play a vital role in visualizing and analyzing the weld pool, observing arc dynamics, capturing droplet transfer, and identifying defects using advanced tools like Digital Image Correlation (DIC). Additionally, acoustic sensors are effective in evaluating process stability and identifying disturbances or defects based on characteristic sound patterns during welding.

To understand mechanical aspects, dynamometers measure axial forces, particularly in processes involving mechanical motion or interaction. Gauss meters detect arc blow effects—important both for maintaining arc stability and ensuring operator safety. Photonic Doppler velocimetry (PDV) systems offer highly precise velocity and displacement measurements, enhancing feedback control in high-speed or critical welds. Pressure sensors, meanwhile, monitor arc force and melt pool dynamics, supporting closed-loop control strategies.

As shown in Figure 4.17, real-time signal analysis (e.g., wavelet transform of current-voltage signals) further enables welding process optimization and defect prevention. Looking forward, adaptive control systems, powered by machine learning and multi-sensor data fusion, are paving the way for autonomous weld defect correction and intelligent process optimization. The integration of these systems into a "digital twin" framework holds the key to achieving fully automated and intelligent welding operations.

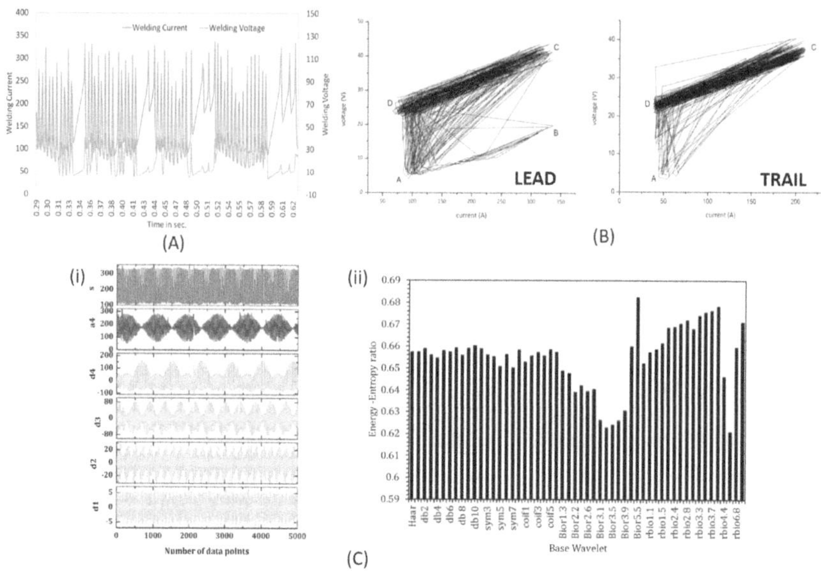

FIGURE 4.17 Signals obtained in pulsed GMAW process: (a) typical current–voltage signal [25], (b) cyclo grams for stability analysis for a case of twin-wire GMAW process [26], and (c) (i) wavelet transformation obtained for typical current–voltage signal and (ii) energy–entropy ratio for different wavelets [26].

REFERENCES

1. Y. Zhang, *Real-time Weld Process Monitoring*. Woodhead Publishing Limited, 2008. https://doi.org/10.1533/9781845694401.
2. V. L. Jorge, F. M. Scotti, R. P. Reis, and A. Scotti, "Wire feed pulsation as a means of inducing surface tension metal transfer in Gas Metal Arc Welding," *J. Manuf. Process.*, vol. 62, pp. 655–669, Jan. 2021. https://doi.org/10.1016/j.jmapro.2020.12.072.
3. A. S. J. S. Panchagnula, S. Goka, and S. Simhambhatla, "Thermal management in manufacture of thin-walled components produced by arc-based additive manufacturing," in *10th International Conference on Trends in Welding Research*. Japan, 2016.
4. J. D. Kulkarni, S. B. Goka, P. K. Parchuri, H. Yamamoto, K. Ito, and S. Simhambhatla, "Microstructure evolution along build direction for thin-wall components fabricated with wire-direct energy deposition," *Rapid Prototyp. J.*, vol. 27, no. 7, pp. 1289–1301, 2021. https://doi.org/10.1108/RPJ-04-2020-0085.
5. A. Kapil, S. Q. Moinuddin, and A. Sharma, "Digitization of welding processes," in *Joining Processes for Dissimilar and Advanced Materials*. Elsevier, 2022, pp. 483–512. https://doi.org/10.1016/B978-0-323-85399-6.00020-5.
6. W. Uhring and M. Zlatanski, "Ultrafast Imaging in Standard (Bi)CMOS technology," in *Photodetectors*. InTech, 2012. https://doi.org/10.5772/35767.
7. Cordin group, "Rotating mirror camera specifications: Model 560," 2020. http://www.cordin.com/index.html.
8. P. W. W. Fuller, "Some Highlights In The History Of High-speed Photography And Photonics As Applied To Ballistics," in *High-Pressure Shock Compression of Solids VIII*, Berlin/Heidelberg: Springer-Verlag, pp. 251–298. https://doi.org/10.1007/3-540-27168-6_7.

9. M. K. Eugene A. Igel, *Rotating Mirror Streak and Framing Cameras*. 1997. https://doi.org/https://spie.org/publications/book/256869#_=_.
10. G. R. Parker, B. W. Asay, and P. M. Dickson, "A technique to capture and compose streak images of explosive events with unpredictable timing," *Rev. Sci. Instrum.*, vol. 81, no. 1, Jan. 2010. https://doi.org/10.1063/1.3284517.
11. C. Xia, Z. Pan, S. Zhang, H. Li, Y. Xu, and S. Chen, "Model-free adaptive iterative learning control of melt pool width in wire arc additive manufacturing," *Int. J. Adv. Manuf. Technol.*, vol. 110, no. 7–8, pp. 2131–2142, 2020. https://doi.org/10.1007/s00170-020-05998-0.
12. S. Goka, M. Manjaiah, and M. J. Davidson, "Study on mechanical, microstructural and corrosion analysis of wire arc additive manufactured AZ31 magnesium alloy," *Met. Mater. Int.*, 2024. https://doi.org/10.1007/s12540-024-01864-w.
13. S. Q. Moinuddin *et al.*, "Research progress in underwater welding: techniques, materials, advancements, and challenges," *Weld. World*, May 2025. https://doi.org/10.1007/s40194-025-02068-5.
14. S. Goka, M. Mallaiah, and M. J. Davidson, "Optimization of bead morphology for cold metal transfer wire arc additive manufacturing of AZ31 magnesium alloy wires," 2024, pp. 129–137. https://doi.org/10.1007/978-3-031-50240-8_27.
15. K. Venkata Rao, "Online modeling and prediction of weld bead geometry in robotic gas metal arc based additive manufacturing using grey prediction model," *Expert Syst. Appl.*, vol. 236, Feb. 2023, p. 121284, 2024. https://doi.org/10.1016/j.eswa.2023.121284.
16. J. Xiong, Y. Zhang, and Y. Pi, "Control of deposition height in WAAM using visual inspection of previous and current layers," *J. Intell. Manuf.*, vol. 32, no. 8, pp. 2209–2217, 2021. https://doi.org/10.1007/s10845-020-01634-6.
17. F. W. C. Farias, J. da Cruz Payão Filho, and V. H. P. Moraes e Oliveira, "Prediction of the interpass temperature of a wire arc additive manufactured wall: FEM simulations and artificial neural network," *Addit. Manuf.*, vol. 48, Oct. 2021. https://doi.org/10.1016/j.addma.2021.102387.
18. C. Xia, Z. Pan, J. Polden, H. Li, Y. Xu, and S. Chen, "Modelling and prediction of surface roughness in wire arc additive manufacturing using machine learning," *J. Intell. Manuf.*, vol. 33, no. 5, pp. 1467–1482, 2022. https://doi.org/10.1007/s10845-020-01725-4.
19. T. Hauser, A. Da Silva, R. T. Reisch, J. Volpp, T. Kamps, and A. F. H. Kaplan, "Fluctuation effects in wire arc additive manufacturing of aluminium analysed by high-speed imaging," *J. Manuf. Process.*, vol. 56, pp. 1088–1098, Aug. 2020. https://doi.org/10.1016/j.jmapro.2020.05.030.
20. J. Xiong and G. Zhang, "Online measurement of bead geometry in GMAW-based additive manufacturing using passive vision," *Meas. Sci. Technol.*, vol. 24, no. 11, p. 115103, Nov. 2013. https://doi.org/10.1088/0957-0233/24/11/115103.
21. R. Yu, Y. Cao, H. Chen, Q. Ye, and Y. M. Zhang, "Deep learning based real-time and in-situ monitoring of weld penetration: Where we are and what are needed revolutionary solutions?," *J. Manuf. Process.*, vol. 93, pp. 15–46, Mar. 2023. https://doi.org/10.1016/j.jmapro.2023.03.011.
22. A. Sharma, A. K. Chaudhary, N. Arora, and B. K. Mishra, "Estimation of heat source model parameters for twin-wire submerged arc welding," *Int. J. Adv. Manuf. Technol.*, vol. 45, no. 11–12, pp. 1096–1103, 2009. https://doi.org/10.1007/s00170-009-2046-3.
23. A. Sharma, N. Arora, and B. K. Mishra, "Mathematical model of bead profile in high deposition welds," *J. Mater. Process. Technol.*, vol. 220, pp. 65–75, 2015. https://doi.org/10.1016/j.jmatprotec.2015.01.009.
24. U. K. Mohanty *et al.*, "Thermal modelling of alternating current square waveform arc welding," *Case Stud. Therm. Eng.*, vol. 25, p. 100885, Jan. 2021. https://doi.org/10.1016/j.csite.2021.100885.

25. M. Kumar, S. Quadir Moinuddin, S. S. Kumar, and A. Sharma, "Discrete wavelet analysis of mutually interfering co-existing welding signals in twin-wire robotic welding," *J. Manuf. Process.*, vol. 63, pp. 139–151, Mar. 2021. https://doi.org/10.1016/j.jmapro.2020.04.048.
26. S. Q. Moinuddin and A. Sharma, "Arc stability and its impact on weld properties and microstructure in anti-phase synchronised synergic-pulsed twin-wire gas metal arc welding," *Mater. Des.*, vol. 67, pp. 293–302, Feb. 2015. https://doi.org/10.1016/j.matdes.2014.11.052.

5 Power Sources for GMAW

5.1 ADVANCEMENTS IN POWER SOURCES

Gas metal arc welding (GMAW), usually referred to as MIG (metal inert gas) welding, has typically operated with traditional power sources that supply the essential current to welding with an arc. These traditional power sources are majorly dependent on constant voltage (CV) models, through maintaining a stable arc length by modifying the current to variations in the arc gap. The following is a detailed description of these power sources:

5.1.1 Transformer-Based Power Sources

Transformer-based systems were the first generation of GMAW power sources. They convert high-voltage, low-current AC from the power supply into a low-voltage, high-current output suitable for welding [1]. These systems are:

- Simple and Consistent: Transformer-based power supplies are known for their durability and reliability.
- Economic: Because of their relatively less complex build, they typically cost less than the more advanced systems.

Nevertheless, these systems have substantial limitations:

- Large and Hefty: Because of the size of the transformers and additional components, these systems are usually large and hard to move, restraining their flexibility.
- Restricted Control: They do not offer accurate control over the welding parameters, which makes it difficult to adjust the arc characteristics or regulate for definite material or thickness needs.

5.1.2 Rectifier-Based Power Sources

With progressing technology, rectifier-based power sources have become the standard for GMAW. These systems convert AC to DC, which provides a refined and steadier arc, vital for reliable weld quality in GMAW. These sources offer:

DOI: 10.1201/9781003778844-5

- Improved Arc Stability: DC output provides a more stable and controllable arc compared to AC, which is vital for high-quality welds.
- Improved Versatility: DC power has the capability for better control during the welding process, which includes enhanced handling of flimsy materials and a wider choice of applications.

Though rectifier-based power sources (see Figure 5.1) characterized an advancement, they shared identical limitations as transformers:

- Bulky: Although an upgrade over transformers, rectifier-based systems were still comparatively large and heavy.
- Limited Flexibility: Though the controls were considerably more innovative than transformer-based systems, they still presented limited adaptability to instantaneous changes in welding conditions.

5.1.3 Motor-Generator Power Sources

Motor-generator systems were another conventional power source for GMAW. These systems make use of an electric motor to drive a generator, producing welding current [2]. These systems offered some flexibility, as they could be powered by several sources of fuel and were frequently used in environments that had no access to electricity. But, motor-generator systems were:

- Not Efficient: These systems used up huge amounts of energy, rendering them costly to operate.
- Unjustified Maintenance: They needed unjustified maintenance because of the mechanical components that were involved.

5.1.4 Developments in Power Sources for GMAW

In recent decades, technological advancements have significantly improved GMAW power sources. These new technologies provide more precise control, greater energy efficiency, and enhanced flexibility. Here are the key advancements:

(A) **Inverter-Based Power Sources**

The introduction of inverter-based power sources (see Figure 5.2) represents a major leap forward for GMAW. Inverters use high-frequency electronics to convert AC into the desired welding current. These systems offer several advantages:

- Compact and Lightweight: Inverters are much smaller and lighter than conventional transformer or rectifier-based systems, making them more portable and easier to use in a variety of environments, including confined spaces and job sites requiring mobility.

- High Efficiency: Inverter-based power sources are much more energy-efficient, converting a higher percentage of electrical energy into welding power and reducing operational costs [3].
- Precise Control: Inverters allow for precise control over welding parameters, including current, voltage, and arc characteristics. This makes it easier to adjust settings based on the type of material, thickness, and desired weld profile, resulting in higher-quality welds with fewer defects.
- Versatile: These power sources can be adapted to a wider range of welding processes, including pulsed MIG welding, which allows for better control over heat input, reducing the risk of warping and improving performance on thinner materials.

Advantages over Conventional Sources:

- Energy Savings: Inverter-based systems can be up to 30% more energy efficient than conventional power sources.
- Flexibility: Inverters can be tuned to work with different materials and welding processes, making them highly adaptable for different welding jobs.
- Portability: Their reduced size and weight compared to transformer systems make them ideal for fieldwork and mobile applications.

(B) **Digital Control and Microprocessor Integration**

Modern GMAW power sources are now equipped with advanced digital controls and microprocessor technology. This allows for real-time monitoring and adjustment of the welding process, optimizing performance and reducing human error. Some features include:

- Adaptive Controls: The system can automatically adjust welding parameters based on the changing conditions of the arc or material, ensuring consistent weld quality.
- Data Logging and Analysis: Many modern power sources can log performance data, allowing welders and engineers to analyze and optimize welding operations.
- Pre-programmed Settings: Welders can select from pre-set programs tailored for specific materials or welding tasks, reducing the need for manual adjustments and increasing productivity.

Advantages over Conventional Sources:

- Improved Consistency: Automated control helps maintain consistent weld quality, even as environmental or material conditions change.

- Increased Productivity: By reducing the need for manual adjustments and fine-tuning, digital control systems can speed up the welding process and reduce downtime.
- Data-Driven Improvements: The ability to collect and analyze data helps companies improve processes, reduce waste, and optimize weld parameters for cost savings.

(C) **Pulsed MIG and Advanced Waveform Control**

Pulsed MIG welding has become increasingly common with the advent of advanced power sources. This technique alternates between high and low current pulses (see Figure 5.3), allowing for better control overheat input and reducing the risk of defects such as spatter and distortion. Modern power sources also allow for advanced waveform control, which can further refine the welding process.

- Reduced Heat Input: Pulsed MIG allows for lower overall heat input, which helps avoid warping and damage to thinner materials [4].
- Improved Control over Weld Pool: The pulsing action gives welders greater control over the weld pool, making it easier to manage complex or out-of-position welds.

Advantages over Conventional Sources:

- Better Weld Quality: Pulsed MIG and waveform control reduce spatter, improve bead appearance, and reduce the likelihood of defects such as burn-through or lack of fusion.
- Expanded Material Range: These technologies make it easier to weld challenging materials such as aluminum and stainless steel, which require precise heat control.

(D) **Automation and Robotic Integration**

The rise of automation and robotics in GMAW has also been supported by advancements in power sources. Modern systems are designed to integrate seamlessly with robotic welding arms, providing the consistent, high-quality welds required for automated production lines.

- Precision Welding: Robotic systems combined with advanced power sources can perform highly precise welds, reducing the likelihood of errors and defects.
- Increased Productivity: Automation significantly increases the speed of welding operations, reducing production times and costs.

Advantages over Conventional Sources:

- Consistency: Robotic systems eliminate human error, ensuring every weld is of the same high quality.
- Efficiency: Automated systems can run continuously, leading to higher output and less downtime.

The advancements in GMAW power sources have significantly improved the welding process. Inverter-based systems, digital controls, pulsed MIG, and robotic integration offer better control, efficiency, and weld quality compared to conventional transformer and rectifier-based power sources. These innovations have expanded the range of materials that can be welded, reduced energy consumption, and improved overall productivity. As GMAW continues to evolve, these advancements set the stage for even greater improvements in automation, process control, and efficiency.

5.2 THE FLEXIBILITY OF MODERN GMAW POWER SOURCES

The flexibility of modern GMAW (gas metal arc welding) power sources is one of the most significant advancements over conventional systems. The ability to adjust, fine-tune, and adapt to a wide variety of welding applications, materials, and environments sets the new generation of power sources apart. The following is a detailed explanation of the flexibility of different advanced power sources:

5.2.1 Inverter-Based Power Sources

Inverter-based power sources have revolutionized GMAW due to their compact size, precise control, and energy efficiency. Their flexibility is manifested in several ways:

a. **Portability and Compactness**

- Small and Lightweight: Unlike the bulky transformer-based power sources, inverter-based systems are much smaller and lighter. This makes them easier to transport to job sites, particularly for applications in fieldwork or remote locations. For welders who need mobility, these systems offer unparalleled flexibility [5].
- Ideal for Confined Spaces: In industries like shipbuilding or construction, where space is limited, the compact size of inverter-based systems allows welders to work more efficiently without the need for large, stationary equipment.

b. **Wide Range of Material Compatibility**

- Adjustable Output: Inverters provide precise control over current and voltage settings, making them adaptable to a wide range of materials. Whether welding thin materials like aluminum or thick steel plates, an inverter power source can be finely tuned to deliver the exact arc characteristics required for the job.

- Multi-process Capabilities: Many inverter-based systems are capable of handling different welding processes like MIG, TIG, and stick welding, providing further versatility. This adaptability makes it possible for welders to switch between processes without needing a different machine.

c. **Variable Input Power**

- Auto-switching: Inverter-based power sources can automatically adjust to different input voltages (e.g., 110 V, 220 V, or even three-phase power). This allows the machine to be used in various environments without requiring specialized power infrastructure.
- Energy Efficiency: These systems are highly efficient, using less power than traditional transformer systems. In cases where electricity supply is limited or where energy costs are a concern, the flexibility of inverter-based machines in optimizing energy use is a major advantage.

d. **Fine Control Over Arc Characteristics**

- Precise Arc Control: Inverters can adjust arc characteristics in real time, which is particularly important for challenging applications like out-of-position welding or working with materials that are difficult to weld. The ability to control factors like arc length, heat input, and travel speed provides welders with greater flexibility in achieving optimal results.
- Rapid Response to Changes: The digital electronics in inverter power sources enable fast adjustments to any changes in the arc conditions, such as varying arc lengths or sudden shifts in heat input requirements.

5.2.2 Digital Control and Microprocessor Integration

The integration of digital controls and microprocessors in modern GMAW power sources introduces a whole new level of flexibility, enabling precise management of the welding process.

a. **Adaptive Controls**
   - Real-Time Monitoring and Adjustment: Digital controls allow for the automatic adjustment of welding parameters based on the real-time conditions of the arc and the material being welded [6]. For example, if the arc length changes, the system can automatically adjust the current or voltage to maintain stability. This makes it easier for welders to handle inconsistent materials or environmental conditions.
   - Pre-programmed Settings: Many modern systems come with built-in programs for specific welding applications, materials, or thicknesses. A welder can select a pre-programmed setting, and the machine automatically adjusts to provide the ideal conditions for that specific task. This feature is highly flexible, reducing the need for manual adjustments and fine-tuning.

b. **Customizable Welding Profiles**
   - Stored Welding Profiles: Welders can store custom settings for frequently performed tasks. This allows for faster setup times and ensures consistent weld quality. For example, if a welder is repeatedly working on the same material and thickness, they can recall the settings without manually reconfiguring the machine each time.
   - Fine-Tuning Capability: With digital controls, welders have the flexibility to adjust not only current and voltage but also advanced parameters like pulse frequency, arc force, and inductance. These fine adjustments enable superior control over the weld pool, resulting in better weld quality across a wide variety of materials.

c. **Remote Monitoring and Data Logging**
   - Remote Access: Some advanced power sources allow for remote monitoring of welding operations via a network connection. This provides flexibility for supervisors to monitor performance from a distance and adjust if necessary.
   - Data-Driven Flexibility: By logging data, welders and engineers can analyze the performance of the welding process and make data-driven adjustments to improve efficiency, reduce waste, and optimize settings for future tasks. This offers a level of flexibility that wasn't possible with conventional systems, enabling continuous improvement over time.

5.2.3 Pulsed MIG and Advanced Waveform Control

Pulsed MIG welding and advanced waveform control have become game-changers in modern GMAW systems, providing welders with unprecedented flexibility in managing heat input and controlling the arc.

a. **Pulse Control for Different Materials**
   - Heat-Sensitive Materials: Pulsed MIG allows welders to handle heat-sensitive materials like aluminum, stainless steel, and thin-gauge metals. By alternating between high and low currents (pulses), pulsed MIG reduces heat input, preventing issues like warping, burn-through, or distortion. This flexibility makes it ideal for applications requiring tight control over heat input.
   - Adjustable Pulse Parameters: Welders can fine-tune pulse parameters, such as frequency, amplitude, and duration, to suit different materials or joint designs. This adaptability provides more control over the weld pool and penetration, improving overall weld quality.

b. **Versatility in Weld Positions**
   - Out-of-Position Welding: Pulsed MIG offers better control of the weld pool in out-of-position welding, such as overhead or vertical welding. The flexibility to reduce heat input while maintaining a stable arc is crucial in these situations, as it helps prevent gravity from pulling the weld pool out of position.

c. **Waveform Control for Specific Applications**
    - Tailored Waveforms: Advanced waveform control allows welders to select specific arc profiles for different tasks. For instance, a soft arc might be used for thin materials, while a harsh arc can be selected for deep penetration on thicker materials. This adaptability ensures that the power source can handle a variety of welding challenges.
    - Reduced Spatter and Improved Bead Appearance: Advanced waveform control helps reduce spatter and improve bead aesthetics, which can be crucial in applications where weld appearance is as important as strength, such as in automotive or aerospace manufacturing.

5.2.4 Automation and Robotic Integration

The integration of modern GMAW power sources with robotic welding systems (see Figure 5.4) and automation has introduced a new level of flexibility in industrial welding applications.

a. **Automated Welding Flexibility**
    - Programmable Welding Sequences: Robotic welding systems can be programmed to perform complex welding sequences with minimal human intervention. This provides flexibility in manufacturing environments, allowing companies to switch between different products or welding tasks without needing significant downtime.
    - High Precision: Robotic systems, combined with advanced power sources, offer high precision, ensuring consistent weld quality even for complex geometries. This precision is especially important in industries like automotive and aerospace, where safety and consistency are critical.

b. **Rapid Reconfiguration for Different Tasks**
    - Quick Setup Changes: With the use of programmable power sources, automated systems can be quickly reconfigured to handle different welding tasks. For example, a robotic welder can switch between different materials or joint designs with minimal manual intervention.
    - Improved Productivity: The ability to automate repetitive tasks allows for higher throughput and more consistent output, which is crucial in high-volume production environments.

Summary: The flexibility of modern GMAW power sources, driven by advancements in inverter technology, digital control, pulsed MIG, waveform control, and automation, represents a significant improvement over traditional systems. These advancements allow for:

- Greater control over the welding process, enabling welders to handle a wider range of materials and conditions.

- Portability and versatility make it easier to perform high-quality welds in various environments, from factory floors to remote job sites.
- Energy efficiency and cost savings improve overall productivity while reducing operational costs.
- Integration with automation and robotics, allowing for precise, consistent welding in high-production environments.

In summary, these advancements offer welders and manufacturers the flexibility to adapt to increasingly complex welding challenges, providing a higher level of precision, control, and efficiency than ever before.

5.3 AUTOMATED WELDING SYSTEMS: A DETAILED OVERVIEW

Automated welding systems have revolutionized modern manufacturing, improving efficiency, consistency, and safety across industries like automotive, aerospace, shipbuilding, and construction. These systems are powered by robotic arms, automated controllers, and advanced sensors that perform welding tasks with minimal human intervention. The use of automation in welding processes is becoming increasingly prevalent as industries seek to boost productivity while maintaining high levels of precision and quality [7]

This description will explore the components of automated welding systems, the different types of automation, and the benefits and challenges they bring to the manufacturing process.

5.3.1 Components of Automated Welding Systems

Automated welding systems comprise several key components that work together to execute precise welding operations. These include:

1. Robotic Welding Arms: Robotic arms are the core of most automated welding systems. These programmable arms hold and manipulate the welding torch, controlling the motion, speed, and angle of the weld. Robotic arms are equipped with multiple degrees of freedom, allowing them to reach complex angles and complete intricate welds with precision. They can be programmed to repeat the same welding task indefinitely, ensuring consistency across production lines.
2. Power Source: The welding power source provides the necessary electrical energy for the welding process. Automated systems use advanced power sources with digital controls to regulate the voltage, current, and other parameters of the welding arc. Modern power sources can handle multiple welding processes (such as MIG, TIG, and plasma welding) and allow for adaptive controls to adjust to different materials and joint designs.
3. Welding Torch and Consumables: The welding torch in an automated system is responsible for delivering the welding arc and filler material to the weld pool. Depending on the process (e.g., GMAW or GTAW), the torch

may feed a continuous wire electrode or deliver shielding gas to protect the weld from contamination. The torch is connected to the robotic arm and programmed to follow specific welding paths.

4. Sensors and Vision Systems: Sensors are essential for monitoring the welding process in real-time. They provide feedback on parameters like arc length, temperature, and weld quality. Advanced vision systems (see Figure 5.5) can detect defects in the weld seam and automatically adjust the welding parameters to correct any issues. These sensors improve precision and help ensure that the welds meet specified quality standards [8].
5. Control Systems: Programmable logic controllers (PLCs) and computer numerical control (CNC) systems are used to program and control the robotic arm and other components. Control systems allow for the precise execution of complex welding sequences. They also manage the coordination between multiple machines and parts in larger automated welding cells.
6. Workpiece Positioners: Automated welding systems often incorporate positioners or fixtures to hold the workpieces in place. These can rotate or move the workpiece during welding, ensuring that the robot can access all areas of the joint. This is especially useful for welding large or irregularly shaped parts.

5.3.2 Types of Automated Welding Systems

Automated welding can be divided into three primary types: hard automation, soft automation (robotic welding), and hybrid systems. Each type offers different levels of flexibility and is suited to specific applications.

5.3.2.1 Hard Automation (Fixed Automation)

In hard automation systems, machines are designed to perform a specific task in a fixed sequence. This type of automation is ideal for high-volume, repetitive tasks such as seam welding in automotive manufacturing. The machines are custom-built for a particular product, making them highly efficient but less adaptable to changes in design or production.

Advantages: High-speed production, consistent quality.

Disadvantages: Lack of flexibility and high initial setup cost.

5.3.2.2 Soft Automation (Robotic Welding)

Soft automation uses programmable robotic arms to perform welding tasks. These systems are flexible and can be reprogrammed to handle different welding tasks. Robotic welding is widely used in industries like aerospace and automotive, where production lines frequently change and require adaptability.

Advantages: Flexibility, ease of reprogramming, and ability to handle multiple welding tasks with minimal downtime.

Disadvantages: Higher upfront costs and programming complexity compared to manual welding.

5.3.2.3 Hybrid Systems

Hybrid systems combine elements of manual and automated welding. In these systems, an operator may initiate the process and perform more intricate tasks, while the robot handles the repetitive or dangerous aspects of the welding process. This is commonly used in applications where full automation is not feasible due to the complexity of the parts or joints.

Advantages: Combines the precision of automation with human flexibility and oversight.

Disadvantages: Requires skilled operators and may not achieve the same speed as fully automated systems [9].

Advantages of Automated Welding Systems

Automated welding systems offer several significant advantages over manual welding processes:

1. Increased Productivity: Automated systems can work continuously without breaks, significantly increasing the speed of production. Robots can also work at a faster pace than human welders, leading to shorter production times.
2. Improved Consistency and Quality: Automated systems can repeat the same welding task with a high degree of precision, ensuring consistent weld quality across multiple units. This reduces the risk of human error, such as inconsistent arc length or weld pool management.
3. Enhanced Safety: Welding can expose workers to hazardous conditions, including heat, fumes, and sparks. Automated welding systems remove the need for human workers to be in close proximity to the welding arc, reducing the risk of injury and exposure to harmful substances [10].
4. Cost Savings: Although the initial investment in automated systems can be high, the long-term savings in labor costs, reduced material waste, and improved throughput can outweigh the upfront expenses. Robots can operate 24/7, leading to lower per-unit production costs.
5. Adaptability: Robotic welding systems can be reprogrammed to handle different products or tasks, making them highly flexible in industries with changing product designs or varying production volumes.

Challenges of Automated Welding Systems

Despite their many advantages, automated welding systems come with challenges that need to be addressed for successful implementation:

1. High Initial Investment: The upfront cost of purchasing, installing, and programming automated systems can be prohibitive for small manufacturers. The expense includes not only the robots themselves but also the necessary infrastructure, training, and maintenance.
2. Complexity in Setup and Maintenance: Setting up an automated welding system requires skilled personnel who understand programming and

welding techniques. Ongoing maintenance is also necessary to ensure the robots and sensors function correctly, which can add to operational costs.

3. Limited Flexibility in Low-Volume Production: While automated systems are highly efficient for high-volume production, they may not be cost-effective for smaller production runs. In such cases, the time required to program the robots and set up the welding cells may outweigh the benefits of automation.
4. Technical Expertise Requirement: To fully leverage the advantages of automated welding, companies need skilled workers who can program and maintain these systems. This can be a barrier for organizations with limited access to skilled personnel.

Future Trends and Innovations

The future of automated welding is promising, with new technologies expected to further enhance the capabilities of these systems. Artificial intelligence (AI) and machine learning (ML) are being incorporated into welding systems, enabling real-time process optimization and predictive maintenance. Vision systems and sensors are becoming more advanced, providing better feedback and enabling adaptive controls that can respond to changes in the environment or material properties.

The rise of cobots (collaborative robots) is another significant trend. Cobots are designed to work alongside human operators, providing a balance between the flexibility of manual welding and the precision of automation. They are safer and easier to program than traditional robots, making them suitable for smaller businesses looking to introduce automation without a massive upfront investment [11].

Summary: Automated welding systems are transforming the manufacturing landscape by increasing productivity, improving quality, and enhancing safety. With advancements in robotics, control systems, and sensor technology, these systems are becoming more flexible and adaptable to a wide range of welding tasks. However, the challenges of high costs, setup complexity, and the need for skilled personnel must be addressed to fully capitalize on the benefits of automation in welding. As technology continues to evolve, the role of automated welding systems will expand, offering even greater efficiency and precision in the future.

5.3.3 Intelligent Welding Systems: An Overview and Their Benefits over Conventional Welding Systems

The manufacturing world is continually evolving, and welding—a critical process in industries such as construction, automotive, aerospace, and shipbuilding—has not been left behind. With advances in technology, particularly artificial intelligence (AI), machine learning (ML), and data analytics, welding systems are becoming increasingly intelligent. Intelligent welding systems (IWS) go beyond traditional automated or robotic welding systems by integrating smart technology to monitor, analyze, and adapt in real time. These systems are ushering in a new era of precision, efficiency, and flexibility, surpassing the capabilities of conventional welding systems [12].

This description will explore the key features of intelligent welding systems, their benefits compared to conventional welding, and the future they promise for industries relying on welding.

(A) Key Features of Intelligent Welding Systems

Intelligent welding systems (IWS) incorporate advanced technologies that enable them to learn, adapt, and make decisions during the welding process. These systems are typically equipped with sensors, real-time monitoring capabilities, AI-based algorithms, and data analytics tools to optimize the welding operation. Some of the defining features include:

1. Artificial Intelligence and Machine Learning Integration

IWS leverage AI to monitor and control the welding process. Machine learning algorithms allow these systems to analyze vast amounts of welding data in real-time, identify patterns, and make decisions about parameter adjustments such as current, voltage, travel speed, or heat input. The system can "learn" from past welding operations to continuously improve performance, making it more efficient over time.

2. Real-Time Monitoring and Feedback

Unlike conventional systems, intelligent welding systems are equipped with sensors and vision systems that provide real-time feedback on key parameters such as weld pool behavior, arc stability, and joint alignment. This feedback is continuously processed by the system's AI, allowing for real-time adjustments to optimize weld quality.

These sensors can detect potential defects, such as undercut, porosity, or incomplete fusion, and adjust the welding parameters accordingly before a defect even occurs. This predictive control significantly reduces the need for post-weld inspection and rework.

3. Automated Parameter Control

IWS use advanced control systems that automatically adjust welding parameters based on the conditions of the weld and environmental factors such as humidity, temperature, or material thickness. Traditional welding often relies on static presets, but intelligent systems dynamically adjust parameters to optimize efficiency and quality.

This flexibility is particularly useful when welding diverse materials or working in environments where the welding conditions are constantly changing.

4. Data Analytics and Predictive Maintenance

IWS collect and analyze data from multiple welding processes, providing insights into machine performance, weld quality, and potential maintenance needs [13].

With predictive analytics, the system can identify when maintenance is required, preventing unexpected breakdowns and reducing downtime. It can also alert operators to anomalies in the process before they result in faulty welds.

5. **Adaptive Welding Process**

IWS can adapt to different welding processes such as gas metal arc welding (GMAW), Tungsten Inert Gas (TIG), and shielded metal arc welding (SMAW) based on the task at hand. They can also switch between different welding techniques, such as pulsed welding or spray transfer, depending on the material or joint configuration.

(B) **Benefits of Intelligent Welding Systems over Conventional Welding Systems**

Intelligent welding systems offer several advantages over conventional welding systems, which rely more on human input, preset parameters, and basic automation. The introduction of smart technologies transforms the way welding is conducted, leading to superior performance in several key areas:

1. **Increased Efficiency and Productivity**

Automated Decision-Making: In conventional systems, welders or operators must manually adjust parameters during the welding process. With IWS, these adjustments are made automatically in real time. This reduces downtime spent on reconfiguring the machine, leading to faster welding cycles and higher throughput.

Minimized Human Error: One of the most significant efficiency gains comes from reducing human intervention. Human welders may experience fatigue, inconsistency, or mistakes over time, whereas intelligent systems maintain consistent performance, improving overall productivity.

Faster Setup Times: IWS can automatically adjust to different materials or joint designs without needing significant reconfiguration, reducing the setup time between different welding tasks or product lines.

2. **Superior Weld Quality**

Real-Time Adaptive Control: IWS continuously monitor weld conditions and adjust parameters to ensure the highest possible quality. This real-time feedback and control mechanism lead to fewer defects and improved joint strength and appearance compared to conventional welding systems, where operators rely on static settings and visual inspection.

Reduced Rework and Scrap: Conventional welding systems can result in errors that are often not detected until post-weld inspections, leading to costly rework or scrapped parts. IWS prevent such issues by detecting potential problems during the welding process itself and making necessary adjustments.

Consistent Quality Across Complex Welds: Complex joint configurations or difficult-to-weld materials often challenge human welders or conventional automated systems. IWS excel in these scenarios by making precise, data-driven adjustments that ensure consistent quality even in complex or out-of-position welds.

3. **Improved Cost Efficiency**

Material and Energy Savings: IWS reduce wastage by optimizing material use and minimizing defective welds. They also optimize energy consumption by adjusting parameters like arc length and heat input in real time, leading to more efficient energy use compared to conventional welding systems, which may consume excess power due to preset or manually adjusted settings.

- Less Downtime: The predictive maintenance capabilities of intelligent systems help to prevent unexpected breakdowns. By tracking machine performance and scheduling maintenance when necessary, companies can reduce unplanned downtime and extend the lifespan of welding equipment.

4. **Enhanced Safety and Operator Assistance**

Minimized Human Exposure: IWS reduce the need for human welders to be in proximity to the welding arc, which is associated with risks like exposure to fumes, radiation, and extreme heat. This leads to a safer work environment, as human intervention is primarily limited to overseeing the system or performing maintenance tasks.

Augmented Welding Expertise: IWS can assist less experienced welders by providing recommendations or automatically adjusting parameters to ensure high-quality welds. This means that the learning curve for new welders is reduced, and companies can rely on their welding system to maintain quality even with a less experienced workforce.

5. **Adaptability and Flexibility**

Multi-process Capabilities: One of the most significant benefits of IWS over conventional systems is their ability to adapt to different welding processes and materials. This flexibility allows manufacturers to handle a wide range of applications without needing to invest in separate equipment for each welding process.

- Scalability: As industries evolve and product designs change, intelligent welding systems can be easily reprogrammed to accommodate new tasks. Conventional welding systems, on the other hand, may require significant retooling or new hardware.

(D) Challenges and Considerations for Intelligent Welding Systems

Despite their advantages, intelligent welding systems are not without challenges. High initial investment costs, the need for skilled operators to manage and maintain the system, and the complexity of integrating AI into existing manufacturing processes are significant hurdles. However, as technology advances and costs decrease, these systems will likely become more accessible to a broader range of industries.

Additionally, continuous advancements in AI and machine learning will improve the self-learning capabilities of these systems, enabling them to handle more complex welding tasks with even greater precision. For companies that adopt IWS, the return on investment (ROI) is expected to be high due to the efficiency gains and quality improvements [14].

(E) The Future of Intelligent Welding Systems

The future of intelligent welding systems is bright, with continued advancements in AI, machine learning, sensor technology, and robotics. These systems are expected to become even more autonomous, with the ability to learn and improve over time without human intervention. The use of cloud-based data analytics and the Internet of Things (IoT) will also allow for more comprehensive monitoring of welding processes and remote control of welding systems across global manufacturing networks.

Incorporating collaborative robots (cobots), which work alongside human welders, into intelligent welding systems will further improve flexibility and enable the seamless transition between manual and automated processes.

Summary: Intelligent welding systems represent the next frontier in welding technology. By incorporating AI, machine learning, real-time monitoring, and advanced automation, these systems offer significant improvements over conventional welding systems in terms of efficiency, quality, cost savings, and safety. While the initial investment may be high, the long-term benefits of IWS—such as reduced downtime, higher productivity, and superior weld quality—make them a promising solution for industries looking to stay competitive in a rapidly evolving manufacturing landscape. The future of welding will undoubtedly continue to evolve toward smarter, more adaptive systems, transforming how industries approach this critical process.

5.4 CONCLUSION

The transformation of GMAW power sources from basic transformer-driven models to sophisticated inverter-based and digitally controlled systems marks a significant leap in welding technology. These advancements have introduced unprecedented control, adaptability, and efficiency, enabling welders to optimize process parameters in real time for a wide variety of applications. Inverter-based sources offer portability, energy savings, and fine arc control, while digital integration allows for adaptive feedback, stored weld profiles, and remote monitoring.

Beyond flexibility, modern power sources serve as the core enablers of automated welding systems, facilitating precise, repeatable, and high-throughput welds in industrial environments. Further, the emergence of intelligent welding systems, equipped with AI, machine learning, and real-time sensor integration, pushes the boundaries of conventional welding. These systems not only enhance weld quality and productivity but also minimize human intervention and rework through predictive and adaptive controls.

As manufacturing evolves toward digitalization and autonomy, GMAW power sources are becoming smarter, more responsive, and seamlessly integrated into intelligent production lines. Their role extends beyond merely supplying energy—they are becoming the digital backbone of next-generation welding systems.

REFERENCES

1. R. Shiju Kumara and T. Ruban Deva Prakash, Design of gas tungsten arc welding power supply using power electronic transformer for steel and aluminium welding," Int. J. Curr. Eng. Technol., vol. 3, no. 4, pp. 1484–1485, 2013. http://inpressco.com/category/ijcet.
2. D. S. Naidu, S. Ozcelik, and K. L. Moore, "Gas metal arc welding: Modeling," in Modeling, Sensing and Control of Gas Metal Arc Welding. Amsterdam, Netherlands: Elsevier, 2003.
3. C. Li, C. Du, H. Xu, and Y. Luo, "The design and study of GMAW inverter power based on DSCCS," Procedia Eng., vol. 15, pp. 1098–1102, 2011. https://doi.org/10.1016/j.proeng.2011.08.203.
4. E. Street, "Introduction," in Pulsed Arc Welding. Abington: Abington Publishing, 1990, pp. 7–7.
5. G. L. Mukhopadhyay, "Development of inverter welding power sources," Indian Weld. J., vol. 31, no. 3, pp. 26–27, 1998. https://doi.org/10.22486/iwj.v31i3.177621.
6. N. Huang, J. Zhang, T. Zhang, X. Zheng, and Z. Yan, "Control of welding speed and current in machine–human cooperative welding processes," Crystals, vol. 12, p. 235, 2022. https://doi.org/10.3390/cryst12020235.
7. E. A. Gyasi, P. Kah, S. Penttilä, J. Ratava, H. Handroos, and L. Sanbao, "Digitalized automated welding systems for weld quality predictions and reliability," Procedia Manuf., vol. 38, pp. 133–141, 2019. https://doi.org/10.1016/j.promfg.2020.01.018.
8. R. H. G. e. Silva, D. Galeazzi, M. B. Schwedersky, et al., "An adaptive orbital system based on laser vision sensor for pipeline GMAW welding," J. Braz. Soc. Mech. Sci. Eng., vol. 43, p. 358, 2021. https://doi.org/10.1007/s40430-021-03081-7.
9. B. Acherjee, "Hybrid laser arc welding: State-of-art review," Opt. Laser Technol., vol. 99, pp. 60–71, 2018. https://doi.org/10.1016/j.optlastec.2017.09.038.
10. A. Kumar Gupta, V. Kumar Chouksey, and A. Pandey, "Design and study of an autonomous linear welding robot with mechanical referencing system," Mater. Today: Proc., 2023. https://doi.org/10.1016/j.matpr.2023.08.111.
11. F. Ferraguti, V. Villani, and C. Storchi, "MyWelder: A collaborative system for intuitive robot-assisted welding," Mechatronics, vol. 89, p. 102920, 2023. https://doi.org/10.1016/j.mechatronics.2022.102920.
12. B. Wang, S. J. Hu, L. Sun, and T. Freiheit, "Intelligent welding system technologies: State-of-the-art review and perspectives," J. Manuf. Syst., vol. 56, pp. 373–391, 2020. https://doi.org/10.1016/j.jmsy.2020.06.020.

13. M. Pech, J. Vrchota, and J. Bednář, "Predictive maintenance and intelligent sensors in smart factory: Review," Sensors, vol. 21, no. 4, p. 1470, 2021. https://doi.org/10.3390/s21041470.
14. D. Curiel, F. Veiga, A. Suarez, and P. Villanueva, "Advances in robotic welding for metallic materials: Application of inspection, modeling, monitoring and automation techniques," Metals, vol. 13, p. 711, 2023. https://doi.org/10.3390/met13040711.

6 Gas Metal Arc Welding in Metal Joining

Gas metal arc welding is used to join metals in various ways. However, a few key factors must be considered for weld selection to ensure strength, efficiency, and quality in fabrication. The best weld option depends on the joint design (butt/lap/T-joint), material thickness (groove welds for thicker metals), load type (fillet for shear stress/groove for tension), welding position (flat/vertical/overhead), and shielding gas (CO_2 for penetration/argon mix for cleaner welds).

6.1 TYPE AND POSITION OF THE JOINTS

GMAW is also referred to as MIG (metal inert gas) welding and is a flexible technique for joining different metals. A joint's type is determined by its application, material thickness, and structural needs. The thickness of the material, the required load, and the welding efficiency all play a role in selecting the best GMAW joint type. In structural applications, butt and T-joints are most frequently utilized, while sheet metal is joined by lap and edge joints. Strong, flawless welds are guaranteed by using the right welding technique and joint preparation. There are five types of joints as follows.

6.1.1 Types of Weld Joints

The weld joints are of five types, namely butt joint, lap joint, T-joint, corner joint, and edge joint. When two workpieces are welded together along their edges and oriented in the same plane, a butt joint is created. Common varieties include the single V-groove butt joint, which has beveled edges for thicker metals, the double V-groove butt joint, which is made for very thick materials and requires welding from both sides, and the square groove butt joint, which is used for thin materials without edge preparation. Because of its strength and adaptability, this joint type is frequently utilized in structural steel applications, sheet metal production, and pipe welding.

Lap joints are made by overlapping two workpieces and welding along the edges where they meet. They are very suitable for thin sheets and can be assembled quickly, but they tend to concentrate stress at the weld edges. Because of their simplicity and effectiveness, lap joints are frequently used in shipbuilding, storage tanks, and automotive panels.

Because of its strength and versatility, the T-joint is used extensively in frames, brackets, and structural supports. It is made when two workpieces are positioned perpendicular to one another in a "T" shape. Common variations include the fillet

DOI: 10.1201/9781003778844-6

weld, which requires no edge preparation and is perfect for general use, and the beveled T-joint, which is used for thicker materials to ensure deeper penetration.

Two workpieces meet at a 90° angle to form a corner joint, which results in an "L" shape. The two main types of corner joints are open corner joints, which are external welds for thin sheets, and closed corner joints, which are internal and external welds for thick materials. These joints are frequently used in furniture, sheet metal enclosures, and box fabrication for neat, reliable right-angle connections.

An edge joint welds two parallel workpieces along their edges. Although it is less strong than butt or lap joints, an edge junction, which is frequently used for sheet metal seams, flanges, and light-duty assemblies, offers a straightforward solution for thin materials where deep penetration is not essential. It does this by welding two parallel workpieces along their edges. Figure 6.1 shows various types of weld joints used in the GMAW process.

In GMAW, "weld type" describes the form and technique used to deposit the weld metal into a joint. These are the five main types of welds:

(1) Triangle Fillet Weld—no preparation required for lap/T-joints;
(2) Groove Weld—fills butt joint gaps (square, V, or U groove for different thicknesses);
(3) Plug Weld—fills lap joint holes for localised strength;
(4) Spot Weld—quick, circular fusion for thin sheets; and
(5) Seam Weld—continuous spots for leak-proof seams. Uses include fuel tanks, pipes, vehicle panels, and structural frameworks.

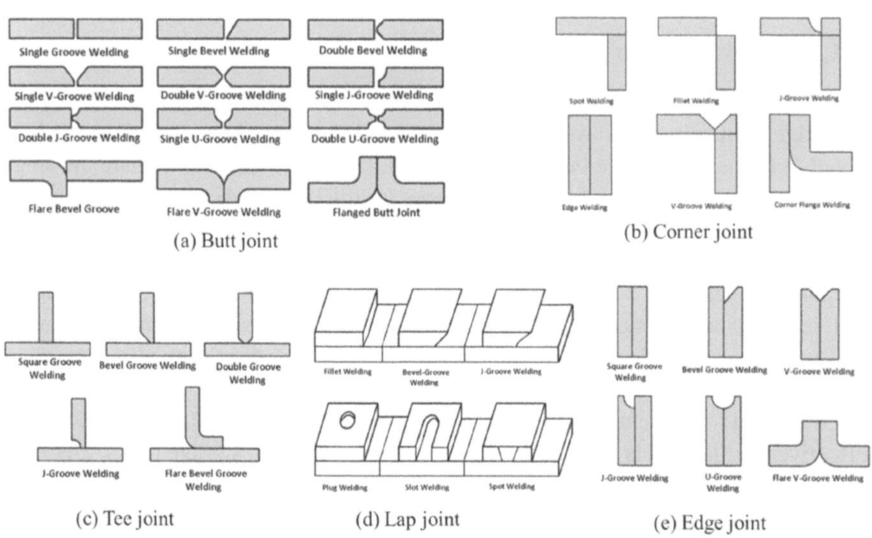

FIGURE 6.1 Types of weld joints [1].

Other specialized weld types:

> When joining rounded edges, like when joining a pipe to a flange, flare-bevel groove welds are utilized. To guarantee complete penetration and reinforce the connection, backing welds are performed to the root side of a groove weld. Conversely, surfacing welds create cladding or wear-resistant coatings rather than joining pieces.

6.1.2 Types of Weld Positions

They are classified into various types depending on the position of the weld joint. As per ISO and AWS, weld positions are categorized 1G–6G [2]. Figure 6.2 briefs the various types of weld positions. They are explained in detail as follows.

6.1.2.1 Flat Position (1G/1F)

The top surface of a horizontal workpiece is welded when the flat welding position is used. It's the simplest and most effective posture with great penetration and little splatter. This technique is perfect for thick plates, bench welding, and structural fabrication.

6.1.2.2 Horizontal Position (2G/2F)

When welding horizontally, as in butt or T-joints, the workpiece stands vertically and the weld axis is horizontal. Because molten metal tends to sag, this configuration can be challenging and requires careful control of the welding gun. Applications such as shipbuilding and pipeline welding frequently employ it.

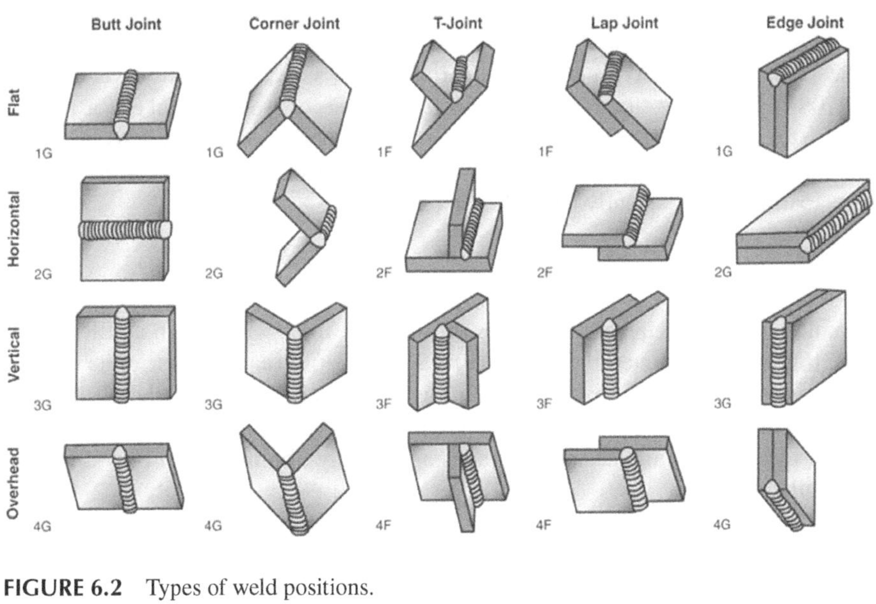

FIGURE 6.2 Types of weld positions.

6.1.2.3 Vertical Position (3G/3F)

There are two methods for vertical welding: Uphill (Vertical-Up), which is slower but more effective for thick metals, and Downhill (Vertical-Down), which is quicker and better suited for thin sheets but has less penetration. It is frequently utilized to construct storage tanks and structural steel projects.

6.1.2.4 Overhead Position (4G/4F)

This is one of the hardest places to weld from since it requires welding from underneath the joint. Because molten metal droops due to gravity, it must be carefully controlled using short arcs and little heat input. Automotive underbody repairs and ceiling welding are its usual applications.

6.1.2.5 Other Positions Used in Pipe Welding are 5G and 6G

In the 5G position, the welder moves around the pipe, which is fixed horizontally. The most difficult and real test of a welder's abilities is the 6G position, which positions the pipe at a 45° angle. These jobs are frequently utilized in the construction of power plants and pipelines for gas and oil.

6.2 SIMILAR METALS

In GMAW, "similar metals" refers to instances where the base material and the filler wire have the same or extremely similar chemical composition. Stronger weld connections, consistent mechanical qualities, and improved metallurgical compatibility are all guaranteed when similar metals are used. A thorough explanation of common, comparable metals used in GMAW is provided below:

i. Carbon Steels

The most used GMAW material is carbon steel, which is usually welded with ER70S-6 filler wire. Its availability, cost, and ease of welding make it a popular choice for structural fabrication, automotive, and construction applications. Although carbon steel is strong and ductile, it has a modest resistance to corrosion and protective coatings are frequently needed for longevity.

ii. Stainless Steels

Stainless steel is frequently welded with matched filler wires, such as ER308L or ER316L, and base materials, such as 304, 308, or 316 series. Because of its superior corrosion resistance and hygienic look, it is extensively utilized in chemical plants, medical devices, and food processing. High levels of chromium and nickel are present in these steels, and austenitic grades are usually non-magnetic, providing both strength and beauty.

iii. Aluminum and Aluminum Alloys

The 5xxx (Al-Mg) and 6xxx (Al-Mg-Si) series alloys with fillers like ER4045, ER5356, or ER4043 are frequently used for welding aluminum and its alloys. Because they are lightweight and have strong corrosion resistance, these materials are widely employed in the automotive, marine, and aerospace industries. To guarantee strong, flawless welds, aluminum must be thoroughly cleaned before welding due to its high heat conductivity.

iv. Nickel Alloys

Matching filler metals like ERNiCrMo-3 are commonly used in the welding of nickel alloys like Inconel 625 or 718. Because of their exceptionally high-temperature strength and resistance to corrosion, these alloys are crucial to the nuclear, power-generating, aerospace, and chemical processing industries. They are perfect for demanding applications since they retain their good mechanical properties even in challenging conditions.

v. Copper and Copper Alloys

Copper and copper alloys, including pure copper and bronze (such as silicon bronze, or ERCuSi-A), are frequently utilized in plumbing, heat exchangers, and electrical contacts. They have good ductility, corrosion resistance, and outstanding electrical and thermal conductivity. Preheating is frequently required to obtain appropriate weld quality because of their high thermal conductivity.

Benefits of using similar metals in GMAW:

> Metallurgical compatibility, which lessens cracking and phase mismatch, and mechanical homogeneity, which guarantees constant strength, hardness, and ductility throughout the weld, are two advantages of using similar metals in GMAW. It is also easier to design and qualify for structural applications when similar metals are used because of their predictable performance and less distortion from matching thermal expansion and contraction

6.2.1 Metallurgical Aspects

Gas metal arc welding is commonly used for the manufacture of similar weld joints (in case of base metals of the same or similar grade). Metallurgical characteristics of similar weld joints in GMAW are welding parameter interaction, base metal, filler metal, and shielding gas characteristics, which influence microstructure, mechanical characteristics, and weld bead quality. The following explores the characteristics of the metallurgical aspects. Besides, the microstructure development and mechanical performance of dissimilar and similar metal CMT-welded joints and material compatibility, joint design, and performance under varying conditions have been discussed.

Anisotropic behavior in mechanical components occurs when manufactured by traditional methods such as casting, forging, and rolling. In the last decade, the anisotropic behavior in mechanical components has been controlled with the help of various techniques, especially with the optimization of process conditions. However, complex geometries would be difficult with preferred modifications of process parameters and in situ operating conditions. To avoid this behavior in complex components, various methods have been suggested.

Metallurgical compatibility is essential to producing high-quality welds when combining the same type of metals using the GMAW process. Dilution effects, phase mismatches, and solidification cracks are greatly reduced since the base and filler metals have comparable compositions. A homogeneous microstructure in the weld zone usually guarantees consistent mechanical qualities like strength and ductility throughout the joint. In the heat-affected zone (HAZ), grain growth is more predictable, and because of matching thermal expansion rates, residual stresses are comparatively smaller. Furthermore, precipitation behavior and solid solution strengthening do not change, leading to stable service performance. All things considered, adopting comparable metals in GMAW enhances joint dependability and streamlines process control.

Figure 6.3(a) shows the microstructure of the base metal (BM) as received from steel. Pearlite and ferrite make up the BM [3]. In Figure 6.3(a), the dark area is ferrite, and the white area is pearlite. The heat-affected zone (HAZ) can be divided into several zones without clear boundaries based on the variation in the peak temperature. Coarse-grained heat-affected zones (CGHAZ), fine-grained heat-affected zones (FGHAZ), inter-critical grain heat-affected zones (ICHAZ), BM from the inner to the outer, and fusion zone (FZ) and weld interface (WI) make up the majority of welded joints. Due to the columnar structure of FZ, the dendrites are primarily oriented toward the weld center and perpendicular to maximize heat transfer. The FZ close to the weld has a honeycomb structure, whilst the weld center itself is made up of isometric dendrites. The temperature differential is enormous, and the growth rate is very slow.

FIGURE 6.3 Optical microscope images for the weld joints, (a) BM, (b) FZ, and (c) HAZ [4].

TABLE 6.1
List of Microstructural and Mechanical Characterization of CMT-Based GMAW Process

Material	Microstructural Properties	Mechanical Properties	References
Magnesium alloy (AZ31)	Equiaxed fine grains of ~ 13.5 μm size	Improved tensile values; UTM = 285 MPa, *e* ~20%	Goka et al. [6]
AZ31B-AA 6061	Intermetallic compound (IMC) layers with α-Mg, $Mg_{17}Al_{12}$	Improved tensile strength	Wang et al. [7]
Al-galvanized steel	Al-Si phase, a combo of Fe_xAl_y phases	Good joint strength	Cao et al. [8]
Al-LCS	Non-uniform IMC layer with cellular structure	Tensile shear force enhanced	Yang et al. [9]
Ti-Cu	IMC layers of Ti-Cu-Al	Same strengths in all the joints	Cao et al. [10]

In this way, similar metal weld joints will represent microstructure with various grain shapes and sizes, depending on the metal coalescence and heat input conditions. As a result, the mechanical properties may vary for the weld joint accordingly.

6.2.2 Mechanical Properties

Similar metals can be joined using GMAW to guarantee consistent mechanical properties throughout the heat-affected zone (HAZ) and weld, maintaining strength and ductility while improving toughness and fatigue resistance important for parts that are subjected to repeated pressures. Performance is directly impacted by this metallurgical compatibility since a microstructure devoid of defects attains tensile strengths (such as 515–620 MPa in 316L welds) and elongation (35–55%) that are comparable to the base metal. In situations where weld integrity determines long-term structural safety, proper heat control avoids brittleness and ensures dependable load-bearing capacity under cyclic stress. Therefore, process optimization is crucial to preventing contamination or thermally caused flaws that jeopardize these outstanding qualities. Table 6.1 shows the summarized information on the metallurgical and mechanical properties of CMT-welded similar metals.

6.3 DISSIMILAR METALS

The materials most frequently used in MIG welding, which finds applications in everything from home appliances to the aerospace sector, are aluminum, steel, and their alloys. These industries will suffer if the characteristics of such aluminum alloys are improved. One of the most widely used techniques for improving the mechanical qualities of all metals and alloys, especially aluminum alloys, is MIG welding.

6.3.1 Metallurgical Aspects

The most common applications for metal inert gas arc welding (MIG) are in the nuclear reactor, petrochemical, power production, telecommunications, and chemical sectors. One innovative way to reduce weight in the industries above is to combine incompatible materials using the welding technique. The thermal, mechanical, and chemical characteristics of the materials present several challenges for the welding of dissimilar materials, including stress, porosity, surface quality, cracking, and strength loss. The effects of MIG welding on the mechanical characteristics of AA 6063, AA2014, and AA7075 were investigated [5]. Mild steel and aluminum alloys have superior mechanical qualities and are lighter than their bulk counterparts. The microstructure of a material can be examined to offer information on whether the structural parameters meet the requirements. Typically, optical or scanning electron microscopes are used for microstructural inspection to magnify the properties of the material being examined as shown in Figure 6.4. These qualities' scale or quantity can be measured, quantified, and compared to acceptability standards.

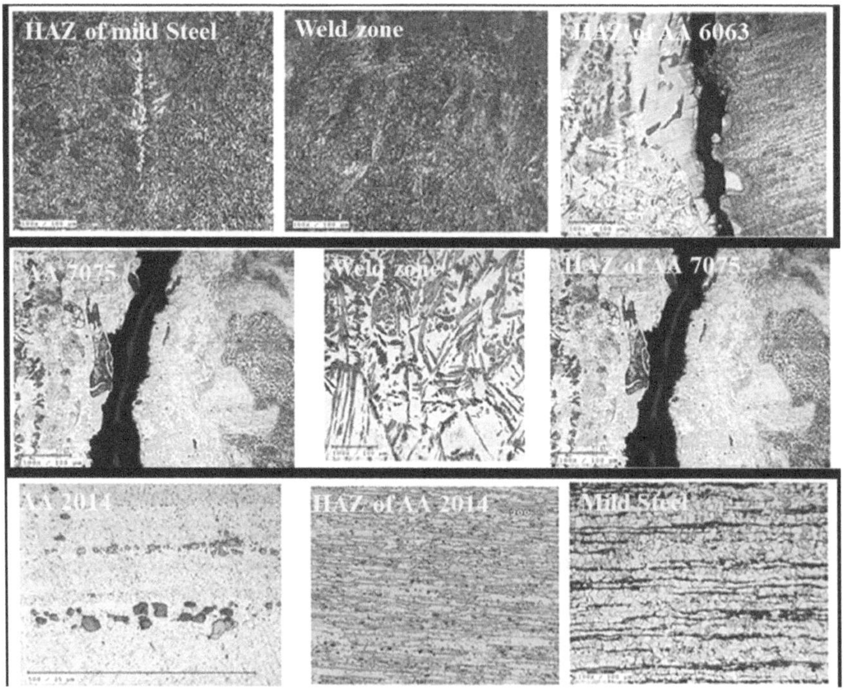

FIGURE 6.4 Microstructure of dissimilar metal joining by GMAW [5].

6.3.2 Mechanical Properties

Table 6.1 shows the extended, summarized information on the metallurgical and mechanical properties of CMT-welded dissimilar metals. It depicts the tensile and grain behavior of similar and dissimilar materials that are subjected to CMT welding.

6.4 JOINING OF NOVEL MATERIALS

Novel materials are emerging to satisfy the need for lightweight, high-performance constructions, such as advanced high-strength steels (AHSS), aluminum alloys, titanium alloys, and composites, such as metal–matrix composites (MMCs). Because of their unique mechanical, thermal, and chemical characteristics, these materials present special welding challenges. The capacity of GMAW to join such materials is essential for improving production processes; nevertheless, to guarantee high-quality welds, process parameters, filler materials, and joint design must be carefully considered.

GMAW welding of innovative materials poses several challenges because of their special characteristics. Because of its high carbon equivalents, AHSS, which is frequently utilized in automotive applications, is susceptible to distortion and cracking as a result of rapid cooling rates. Because of their low melting point and high thermal conductivity, aluminum alloys, which are prized for their lightweight and corrosion resistance, can cause burn-through or partial fusion if not adequately managed. Because titanium alloys, which are essential in aerospace, react strongly with oxygen at high temperatures, strict shielding gas control is required to avoid weld contamination. When welding, metal–matrix composites, which blend metallic and non-metallic phases, frequently experience poor wettability and phase segregation. To overcome these obstacles and produce flawless welds, exact control over the GMAW process parameters is necessary.

The potential to use GMAW to link innovative materials has important ramifications for various sectors. By combining aluminum and AHSS alloys, GMAW makes it possible to produce lightweight, fuel-efficient automobiles. GMAW makes it easier to assemble titanium and composite structures in aerospace, which is essential for low-weight, high-strength parts. The use of GMAW to combine disparate materials for pipelines and renewable energy systems improves the energy sector. In the future, GMAW technology is expected to be better equipped to handle new materials, thanks to developments like robotic automation, real-time process monitoring, and artificial intelligence-driven parameter optimization. Weld quality will be further enhanced by research into novel filler materials and shielding gas compositions designed for certain novel materials. GMAW will continue to play a crucial role in joining innovative materials as industries strive for high-performance and sustainable solutions, spurring innovation in manufacturing techniques.

6.5 MODELING IN MATERIALS JOINING

The process of gas metal arc welding (GMAW) is intricate and involves interactions between heat, metallurgy, and mechanics. Predicting flaws, enhancing mechanical

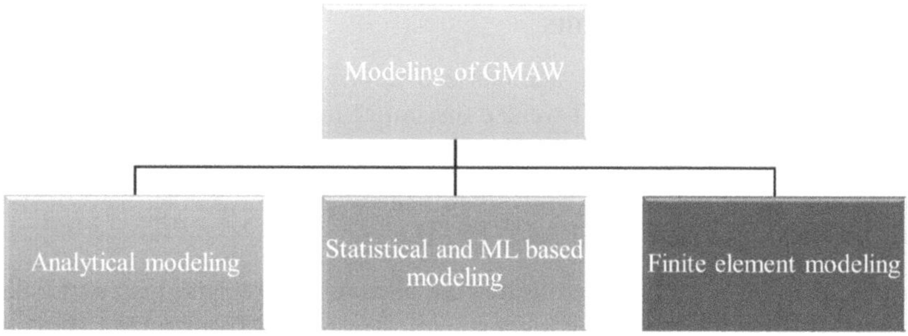

FIGURE 6.5 Types of modeling of the GMAW process.

qualities, and optimizing weld quality all depend on modeling these events. A thorough analysis of the main modeling techniques used in GMAW-based joining is provided below, arranged according to their areas of emphasis.

Figure 6.5 represents various models of the GMAW process. The modeling can analyze microstructure, fluid flow, thermal modeling, residual stress, distortion, etc. Grain structure is predicted using microstructure models (cellular automaton/phase field), heat flow is modeled using thermal analysis (FEA, Goldak model), and molten pool dynamics are modeled using CFD. Thermomechanical finite element analysis is used to simulate residual stresses, and machine learning is used to optimize parameters. Hybrid models address anisotropy and heat accumulation for WAAM. The limits of welding simulation are being pushed by new developments like multiscale modeling and digital twins. To perfect process control and weld quality, each technique is essential.

Modeling of manufacturing processes, such as welding, with the finite element method, is becoming more and more common in engineering practice. Historically, it has mainly been the nuclear industry and the aerospace industry that have been at the forefront of the developments in this area. Safety was the major driving force to ensure the product and process quality. Nowadays, we can see that welding simulations are used more often in other industries. The target may then be to produce a component that is right the first time, a direct hit, by evaluating different design and manufacturing concepts. Thus, cost and time savings are also to be considered. The welding process is a difficult thermo-mechanical process that needs knowledge of various disciplines. The most important relations are shown in Figure 6.6.

In the current research process, a thermo-mechanical analysis is carried out, where the main part of the relations is illustrated and simplified. The current work focuses on the finite element modeling of arc welding, thin wall construction, followed by the corresponding thermo-mechanical analysis.

6.5.1 General Assumptions in the Modeling of Welding

Before building a model for the simulation of welding, a few basic questions have to be clear: what is the purpose of the simulation, how much accuracy is expected,

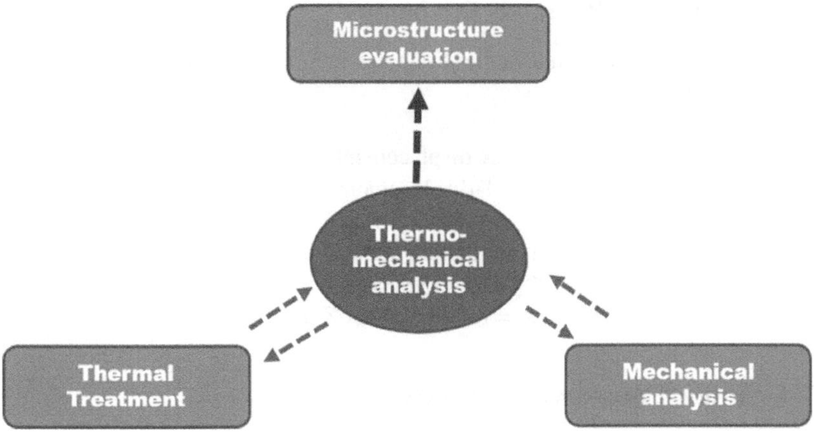

FIGURE 6.6 Disciplines in thermo-mechanical analysis.

and what kind of structure should be observed? Currently, the finite element analysis models have been three dimensional, as the simulations have aimed to validate them for geometries representing the components in engineering practice.

6.5.2 Governing Equations

Three-dimensional Fourier heat conduction transient analysis equation is solved for the modeling of temperature distribution around the heat source.

$$\rho c \frac{\partial T}{\partial t}(x,y,z) = \nabla.\bar{Q}(x,y,z,t) + Q(x,y,z,t) \tag{6.1}$$

Where

ρ = density of the material
c = specific heat capacity
T = current temperature
$\bar{Q}$ = heat flux vector
Q = internal heat generation rate
∇ = special gradient operator

Due to the complete solid domain of our working conditions, the following modified energy equation will be used in the simulations as the governing equation:

$$\rho c \frac{\partial T}{\partial t} = \frac{\partial T}{\partial x}\left(K_T \frac{\partial T}{\partial x}\right) + \frac{\partial T}{\partial y}\left(K_T \frac{\partial T}{\partial y}\right) + \frac{\partial T}{\partial z}\left(K_T \frac{\partial T}{\partial z}\right) \tag{6.2}$$

where K_T is the thermal conductivity in the individual direction (equal in all three directions).

6.5.3 Mechanical Boundary Conditions

Primary mechanical boundary conditions are mentioned concerning displacement and rotation only.

u_x = *X*-displacement
u_y = *Y*-displacement
u_z = *Z*-displacement
Φ_x = rotation about *X*-axis
Φ_y = rotation about *Y*-axis
Φ_z = rotation about *Z*-axis

6.5.4 Thermal Boundary Conditions

Several input parameters have to be added during the thermal analysis to describe the boundary conditions. The initial temperature of the specimen and the ambient temperature play a significant role during the heating and cooling phases. The heat transfer phenomenon is divided into two parts: convection and radiation. The heat transfer could be modeled by different ways in the numerical models. The usually applied opportunities are heat exchange surfaces defined by surface elements or surface loads acting on nodes. The convective heat transfer coefficient is taken equal to 25 W/m²·K. The emissivity coefficient (ε) is assumed to be 0.8, while the ambient temperature (T_{amb}) has to be defined for radiation as well. The radiative heat transfer coefficient is described by

θ_{1} = temperature at point 1
θ_{2} = temperature at point 2
θ_{3} = temperature at point 3
Surface heat flux $q = q(X,t), X \epsilon S$
Volume heat flux $T = r(X,t), X \epsilon \Omega$
Convection $q = h(T - T_0), X \epsilon S$
Radiation $q = (T - T_Z)^4 - (T_0 - T_Z)^4, X \epsilon S$

6.5.5 Creation of the Weld Path

A method has been implemented to navigate the heat source at every time increment. This method elevates the process of tracing the weld path, specifically if the path is an arbitrary 3D line. First, the weld path is defined in a system, e.g., subroutine system, where the tools for handling complex features are generally more developed than in the pre-processors of the finite element systems. Thereafter, the weld path is exported as coordinate points to a file. At the beginning of the simulation, the file with the weld path coordinates is given and processed into the finite element model via user subroutines.

6.5.6 Various Heat Input Models

In a welding simulation phenomenon, the temperature change is translated into thermal expansion in the mechanical analysis, which is the load of the model. Generally, the increase in temperature can be modeled in two different ways: recommended temperatures or by prescribed heat input. When proposing the temperature method, the temperature is given at each node for the particular time and position. Usually, the temperature is linearly ramped to the prescribed value and then followed by a constant temperature phase. This method has been used quite frequently, specifically in the 2D analysis. The other one is the prescribed heat input method, which is the most commonly used method nowadays, spread on the heat input as a heat flux at the integration points that is then transformed to the nodes as temperature loads. The most commonly used heat sources of this kind have a Gaussian distribution, but heat sources with constant distribution have also been used. Interestingly, Goldak et al. [11] proposed a volumetric (double-ellipsoidal shape) heat source, as mentioned in Equations (6.14) and (6.15).

The double-ellipsoidal heat source was initially used for the modeling of GTA-welding process. Furthermore, Goldak et al. [12] addressed that they had received the most accurate temperature field by merging three heat sources: a double elliptical disk on the surface, a double ellipsoid to model the direct impingement of the arc, and another double ellipsoid to model the stirring effect of the molten metal. To model the welds with deep penetration, such as laser beam and electron beam welding, they suggested a conically shaped distribution of power density with a Gaussian distribution radially, in addition to a linear distribution axially. In a study by Lundbäck and Runnemalm [13], a combined double-ellipsoidal heat source and a conical-elliptic heat source are presented. The formula for the frontal part of the conical-elliptic heat source, $z > 0$, can be seen in the equation. There are various types of heat sources used in different types of welding processes. Rosenthal proposed the first moving heat source [14]. Then, many more heat sources were proposed. A few models that are relevant to the present study are discussed in detail.

(a) **Rosenthal Heat Source**

Earlier, Rosenthal proposed a model of moving heat source by using the heat conduction equation of Fourier analysis. The following are the assumptions made by Rosenthal [14, 15]. The author assumed that material properties are independent of temperature, and a point heat source is used in the modeling.
Hence, different equations have been expressed for both 2D and 3D.
Equation for 2D

$$T - T_0 = \frac{q}{2\pi Kt} * e^{-\frac{2\xi}{2\alpha}} * K_0 * \frac{\upsilon R}{2\alpha} \tag{6.3}$$

Equation for 3D

$$T - T_0 = \frac{q}{2\pi Kr} * e^{-\frac{2\xi}{2\alpha}} * \frac{e^{-\frac{\upsilon R}{2\alpha}}}{R} \tag{6.4}$$

Here

$$r=\sqrt{\left(x-x_0\right)^2+\left(y-y_0\right)^2+\left(z-z_0\right)^2}$$

Where

T = bulk temperature (in K)
T_0 = room temperature (in K)
t = plate thickness
K = thermal conductivity
q = total heat (in J/m)
υ = velocity
$\xi = x - vt$ = distance of the point from fixed position (in m)
α = thermal diffusivity (in m/s)
R = sqrt $(x^2 + y^2 + z^2)$ = distance from heat source to particular point
K_0 = Bessel function of first kind, zero order

$$0=z^2\frac{d^2y}{dz^2}+z\frac{dy}{dz}-\left(z^2+v^2\right)y \tag{6.5}$$

where

v = order of the Bessel function
z = argument to the Bessel's function

ii) Adam's solution for peak temperature for 2D heat flow

$$\frac{1}{T_p-T_0}=\frac{1}{T_M-T_0}+\frac{4.3Vyh\rho C}{Q} \tag{6.6}$$

iv) Adam's solution for peak temperature for 3D heat flow

$$\frac{1}{T_p-T_0}=\frac{1}{T_M-T_0}+\frac{5.44\pi k\alpha}{QV}\left(2+\left[\frac{Vy}{2\alpha}\right]^2\right) \tag{6.7}$$

1. **Gaussian Heat source**
a) **Spot Heating**

$$q=\frac{Q}{\pi r^2}*e^{-\left(\frac{(x-x_0)^2+(y-y_0)^2}{r_0^2}\right)} \tag{6.8}$$

where r_0 = radius of the heat source.

i) **Thermal Profile during Spot Heating**

$$T - T_i = \frac{Q}{\rho C \sqrt{\alpha\pi}^3} \int_0^t \frac{1}{\left(4\alpha t + r_0^2\right)\sqrt{t}} \exp\left(\left[-\left(\frac{r^2}{4\alpha t + r_0^2} + \frac{z^2}{4\alpha t}\right)\right] dt\right) \tag{6.9}$$

Here $r = \sqrt{\left(x - x_0\right)^2 + \left(y - y_0\right)^2}$

= Radial distance away from the center of the heat source

b) **Scanning Heating**

$$q = \frac{Q}{\pi r^2} * e^{-\left(\frac{\left(x - x_0 - V_t\right)^2 + \left(y - y_0\right)^2}{r_0^2}\right)} \tag{6.10}$$

Here r_0 = radius heat source.

i) **Thermal Profile during Scanning Heating**

$$T - T_i = \frac{Q}{\rho C \sqrt{\alpha\pi}^3} \int_0^t \frac{1}{\left(4\alpha t + r_0^2\right)\sqrt{t}} exp\left(\left[-\left(\frac{r^2}{4\alpha t + r_0^2} + \frac{z^2}{4\alpha t}\right)\right]\right) dt \tag{6.11}$$

Here $r = \sqrt{\left(x - x_0 - V_t\right)^2 + \left(y - y_0\right)^2}$

= radial distance away from the center of the heat source.

(b) **Pavelic's Disk Heat Source**

Stirring velocity in the weld pool is maximum at the center and reduces to zero at the boundary, as this criterion is not satisfied by the Rosenthal heat source. The concept of a Gaussian distributed heat source is introduced in the modeling of welding. Pavelic et al. [16] defined a 2D disk heat source with a Gaussian heat distribution as shown in Figure 6.7(a).

Heat distribution in the Pavelic heat source is given by the following equation:

$$q(r) = q(0)\exp\left(-cr^2\right) \tag{6.12}$$

Here $q(r)$ = surface flux at any radius from the center of the source (W/m^2),
$q(0)$ = maximum flux (W/m^2) is the concentration coefficient (m^{-2}).

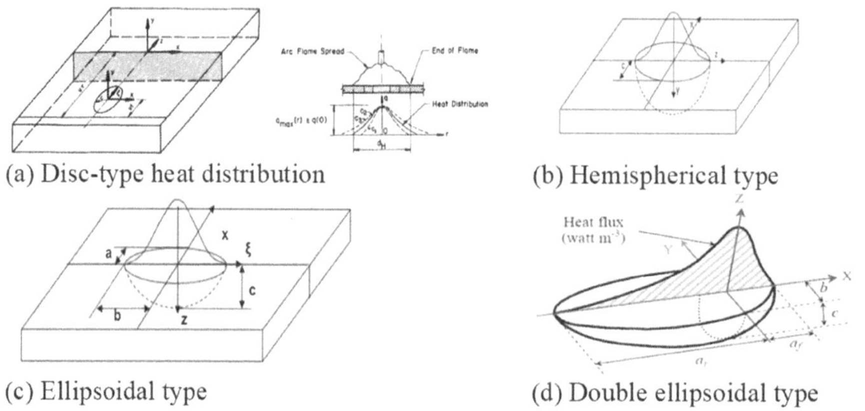

FIGURE 6.7 Heat source models and types [17].

$q(r)$ becomes equal to ηVI when a uniformly distributed heat source of the circular disk $d = \frac{2}{\sqrt{c}}$ is assumed. As the concentration coefficient increases, the diameter of the heat source reduces. Many experimental types of research have shown that some of the heat from the arc is transferred to the base plate in two ways: one directly through molten metal and one more directly to the base plate from convection and radiation. Friedman defined a moving disk heat source by making corrections in Pavelic's equation as follows:

$$Q_g = \left(\frac{3Q}{\pi r_h^2}\right) e^{-3\left(\frac{(x)^2+(\xi)^2}{r_h^2}\right)} \tag{6.13}$$

where

r_h = radius of the disk (m)
Q = total heat supply = ηVI (W)

(c) **Hemispherical Heat Source**

Limitations of disk heat source toward estimation of penetration brought volumetric heat sources into the picture. A review of the hemispherical heat source given by Goldak et al. is as shown in Figure 6.7(b).

Heat distribution in the case of the hemispherical heat source is as shown below:

$$Q_g = \left(\frac{6\sqrt{3}Q}{\pi\sqrt{\pi}c^3}\right) e^{-3\left(\frac{(x)^2+(y-vt)^2+(z)^2}{c^2}\right)} \tag{6.14}$$

where c is the radius of the hemispherical heat source (m).

The shape and size of the weld pool obtained from the hemispherical heat source are far from the realistic weld pool. The hemispherical heat source also has an error regarding the temperature distribution across the fusion region.

(d) **Ellipsoidal Heat Source**

Based on a study of the shape and size of the weld pool, a review of the single-ellipsoidal heat source is shown in Figure 6.7(c). The following is the heat distribution equation in an ellipsoidal heat source.

$$Q_g = \left(\frac{6\sqrt{3}Q}{\pi\sqrt{\pi}\,abc} \right) e^{-3\left(\frac{x^2}{a^2} + \frac{(y-vt)^2}{b^2} + \frac{z^2}{c^2} \right)} \tag{6.15}$$

where Q_g = total heat supplied (W/m³)
a, b, c = heat source parameters in x, y, and z directions, respectively (mm or m).

Comparatively, the temperature prediction using an ellipsoidal heat source is better than that of a hemispherical source. However, in the front part of the heat source, temperature gradients were not as steep as experiment. In the rear part of the heat source, temperature gradients were steeper than the experimental results. Thus, the anomalies led to a double-ellipsoidal heat source as discussed in the next type of heat source.

i. Double-ellipsoidal Goldak's heat source

Initially, Goldak et al. proposed the model of the double-ellipsoidal heat source. This is a more versatile heat source because of its flexibility for various types of processes. Figure 6.7(d) represents the schematic of this source type model. It can be used for shallow depth penetration and deeper penetration welding processes as well. This is the first and one of the effective volumetric heat sources that can be used for non-symmetric welding operations also.

$$Q_f = \left(\frac{6\sqrt{3}Q}{\pi\sqrt{\pi}\,a_f bc} \right) e^{-3\left(\frac{(x-vt)^2}{a_f^2} + \frac{y^2}{b^2} + \frac{z^2}{c^2} \right)} \quad if\ \xi \geq 0 \tag{6.16}$$

$$Q_r = \left(\frac{6\sqrt{3}Q}{\pi\sqrt{\pi}\,a_r bc} \right) e^{-3\left(\frac{(x-vt)^2}{a_r^2} + \frac{y^2}{b^2} + \frac{z^2}{c^2} \right)} \quad if\ \xi < 0 \tag{6.17}$$

where

V = welding velocity (m/s)
Q = total heat input (W)

f_f, f_r = heat fractions supplied in front and rear parts
b, c = heat source parameters in y, z directions, respectively (m)
a_f, a_r = heat source parameters in front and rear parts in the x direction (m)

In the case of Goldak's model, a factor ξ is defined. It is used for moving the coordinate system and is defined as follows:

$$\xi = z + v(\tau - t)$$

where

v = welding velocity in m/s
τ = lag factor used to define the initial position of the heat source
t = current time step in seconds

According to Goldak, the following assumptions are made.

- In the region of the weld pool, high heat transfer is occurring due to shielding gas, stirring velocity, Marangoni effects, and surface tension. To account for this higher heat transfer, he suggested considering a higher thermal conductive coefficient above the melting temperature.
- The length of the heat source in the front part is half the width of the heat source, and in the rear, it is twice the width of the heat source.
- To avoid higher heat loss on the top surface of the heat source due to higher temperature gradients, the top surface of the heat source is considered insulated.
- If the heat source size is 10% less than the weld bead size, then better results are obtained.

Steps in moving the heat source:

- Find the position of the MHS first.
- Next, calculate the affected volume.
- Then find the undertaken nodes of the model that are within the area of heat input.
- Each node should represent some volume. Then, integrate the effect at all the nodes to check the correctness of the heat input

That means, at each node, check the specific heat input (Q_i) and multiply it by the affected volume (V_i).

Figure 6.8 shows a comparison of the outcomes of the simulation and the experiment at significant points during the calculation cycle. The remarkable similarity validated that the temperature and flow field were accurately predicted by the computations. The pulse period droplet moved into the molten pool during the first CMT

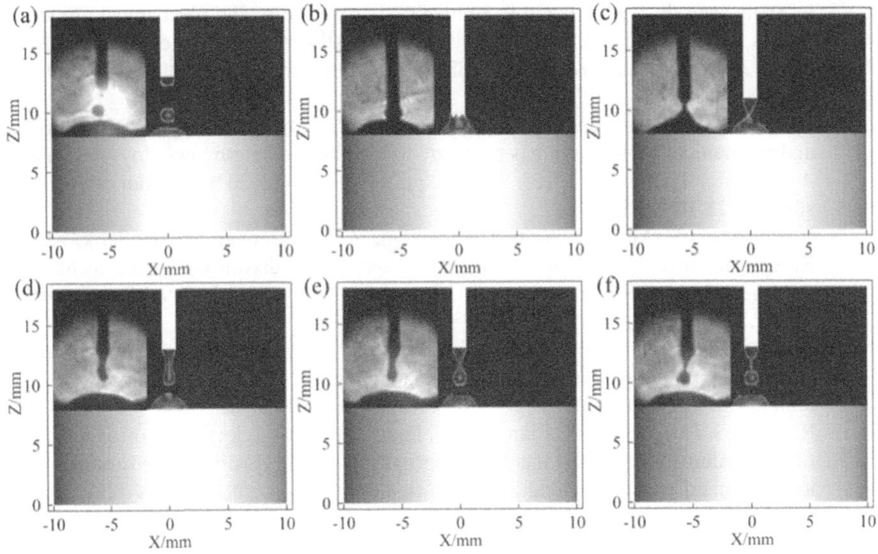

FIGURE 6.8 Comparison of cold metal transfer mechanism through transfer modes (a) $t = 1$ ms, (b) $t = 10$ ms, (c) $t = 15$ ms, (d) $t = 21$ ms, (e) $t = 21.5$ ms, and (f) $t = 22$ ms [18].

period. The droplet finished the short-circuit transfer as the wire fed and retracted. Surface tension, gravity, and the electromagnetic force all worked together to drive the droplet through necking during the next pulse interval.

6.6 CONCLUSIONS

When it comes to combining new materials like metal–matrix composites, titanium alloys, aluminum alloys, and advanced high-strength steels, GMAW is an extremely versatile welding method that is essential despite issues like distortion, cracking, and weld contamination. GMAW produces high-quality welds for these materials by adjusting parameters, using sophisticated processes including cold metal transfer and pulsed GMAW, and using specific filler materials and shielding gases. Its use in the energy, automotive, and aerospace industries highlights how crucial it is to creating lightweight, high-performance structures. GMAW's position in satisfying the changing needs of contemporary production will be further strengthened by upcoming developments in automation, process monitoring, and material-specific advances.

REFERENCES

1. "Types of weld joints," https://www.theengineerspost.com/types-of-welding-joints/.
2. ISO, *Welding and Allied Processes — Symbolic Representation on Drawings*. ISO, 2019.

3. L. Srinivasan, S. J. Jakka, and P. Sathiya, "Microstructure and mechanical properties of Gas tungsten arc welded High Strength Low Alloy (15CDV6) steel joints," *Mater. Today Proc.*, vol. 4, no. 8, pp. 8874–8882, 2017. https://doi.org/10.1016/j.matpr.2017.07.238.
4. D. Zhao, Y. Bezgans, N. Vdonin, L. Radionova, L. Glebov, and V. Bykov, "Metallurgical and mechanical attributes of gas metal arc welded high-strength low-alloy steel," *Int. J. Adv. Manuf. Technol.*, vol. 125, no. 3–4, pp. 1305–1323, 2023. https://doi.org/10.1007/s00170-023-10807-5.
5. P. Sakthivel, V. Manobbala, T. Manikandan, Z. Mohammed Arman Salik, and G. Rajkamal, "Investigation on mechanical properties of dissimilar metals using MIG welding," *Mater. Today Proc.*, vol. 37, pp. 531–536, 2021. https://doi.org/10.1016/j.matpr.2020.05.488.
6. S. Goka, M. Manjaiah, and M. J. Davidson, "Study on mechanical, microstructural and corrosion analysis of wire arc additive manufactured AZ31 magnesium alloy," *Met. Mater. Int.*, 2024. https://doi.org/10.1007/s12540-024-01864-w.
7. P. Wang, S. Hu, J. Shen, Y. Liang, and J. Pang, "Effects of electrode positive/negative ratio on microstructure and mechanical properties of Mg/Al dissimilar variable polarity cold metal transfer welded joints," *Mater. Sci. Eng. A*, vol. 652, pp. 127–135, Jan. 2016. https://doi.org/10.1016/j.msea.2015.11.080.
8. R. Cao, G. Yu, J. H. Chen, and P.-C. Wang, "Cold metal transfer joining aluminum alloys-to-galvanized mild steel," *J. Mater. Process. Technol.*, vol. 213, no. 10, pp. 1753–1763, Oct. 2013. https://doi.org/10.1016/j.jmatprotec.2013.04.004.
9. S. Yang, J. Zhang, J. Lian, and Y. Lei, "Welding of aluminum alloy to zinc coated steel by cold metal transfer," *Mater. Des.*, vol. 49, pp. 602–612, Aug. 2013. https://doi.org/10.1016/j.matdes.2013.01.045.
10. R. Cao, Z. Feng, and J. H. Chen, "Microstructures and properties of titanium–copper lap welded joints by cold metal transfer technology," *Mater. Des.*, vol. 53, pp. 192–201, Jan. 2014. https://doi.org/10.1016/j.matdes.2013.06.030.
11. J. Goldak, A. Chakravarti, and M. Bibby, "A new finite element model for welding heat sources," *Metall. Trans. B*, vol. 15, no. 2, pp. 299–305, Jun. 1984. https://doi.org/10.1007/BF02667333.
12. J. Goldak, M. Bibby, J. Moore, R. House, and B. Patel, "Computer modeling of heat flow in welds," *Metall. Trans. B*, vol. 17, no. 3, pp. 587–600, Sep. 1986. https://doi.org/10.1007/BF02670226.
13. A. Lundbäck and H. Runnemalm, "Validation of three-dimensional finite element model for electron beam welding of Inconel 718," *Sci. Technol. Weld. Join.*, vol. 10, no. 6, pp. 717–724, Dec. 2005. https://doi.org/10.1179/174329305X48347.
14. D. Rosenthal, "The theory of moving sources of heat and its application to metal treatments," *J. Fluids Eng.*, vol. 68, no. 8, pp. 849–865, Nov. 1946. https://doi.org/10.1115/1.4018624.
15. J. A. Ramos-Grez and M. Sen, "Analytical, quasi-stationary Wilson-Rosenthal solution for moving heat sources," *Int. J. Therm. Sci.*, vol. 140, pp. 455–465, Jun. 2019. https://doi.org/10.1016/j.ijthermalsci.2019.03.017.
16. V. Pavelic, R. Tanbakuchi, and O. A. Uyehara, "Experimental and computed temperature histories in gas tungsten-arc welding of thin plates," *Weld. J.*, vol. 48, no. 7, 1969.
17. T. Kik, "Calibration of heat source models in numerical simulations of welding processes," *Metals (Basel).*, vol. 14, no. 11, p. 1213, Oct. 2024. https://doi.org/10.3390/met14111213.
18. Q. Gou, Z. Zhang, L. Xu, D. Wu, T. Zhang, and H. Liu, "Heat and mass transfer behavior in CMT plus pulse arc manufacturing," *Int. J. Mech. Sci.*, vol. 281, p. 109638, Jan. 2024. https://doi.org/10.1016/j.ijmecsci.2024.109638.

7 GMAW in Additive Manufacturing

7.1 WHAT IS AM

AM stands for additive manufacturing, a method of building three-dimensional objects by layering materials according to a computer model. AM allows for more complex geometries and less waste than traditional subtractive manufacturing, which involves removing material from a solid block. Standard methods used in the automotive, medical, and aerospace industries include wire arc additive manufacturing (GMAW in AM), powder bed fusion, and 3D printing. It is essential to investigate the classification and categorization, as modeling aids in process optimization, and mechanical and metallurgical characteristics ensure that the final components fulfill strict requirements for strength and material integrity. Benefits such as cost-effectiveness and intricate geometries are highlighted in WAAM's advantages, which are weighed against drawbacks like potential flaws and post-processing requirements.

7.2 CLASSIFICATIONS OF METAL ADDITIVE MANUFACTURING

Figure 7.1 illustrates a comprehensive classification diagram of metal additive manufacturing (AM) processes, demonstrating various methods for fabricating metal parts in a layer upon layer manner. To represent the variety of methods used in metal additive manufacturing (MAM), it categorizes these techniques into seven major families, each with its subclasses. The shape of the material used (wire, powder, etc.), the types of heating sources (laser, arc, electron beam, etc.), the build volume, and other parameters are some of the factors used to categorize these procedures. The classifications above allow metallic additive manufacturing processes to be separated into various techniques, including SLA, SLS, SLM, WAAM, and others [1].

7.3 WAAM PROCESS

Over the last decade, wire arc additive manufacturing (WAAM), a type of direct energy deposition (DED) technology, has gained recognition as a leading additive manufacturing (AM) method. Classical arc welding technology is the source of WAAM. The wire is heated using an electric arc to melt it in the WAAM process, which moves it to a molten metal pool. Subsequently, it forms a portion layer by layer at the melt pool boundary. Figure 7.2 depicts a simple mechanism of the WAAM process.

DOI: 10.1201/9781003778844-7

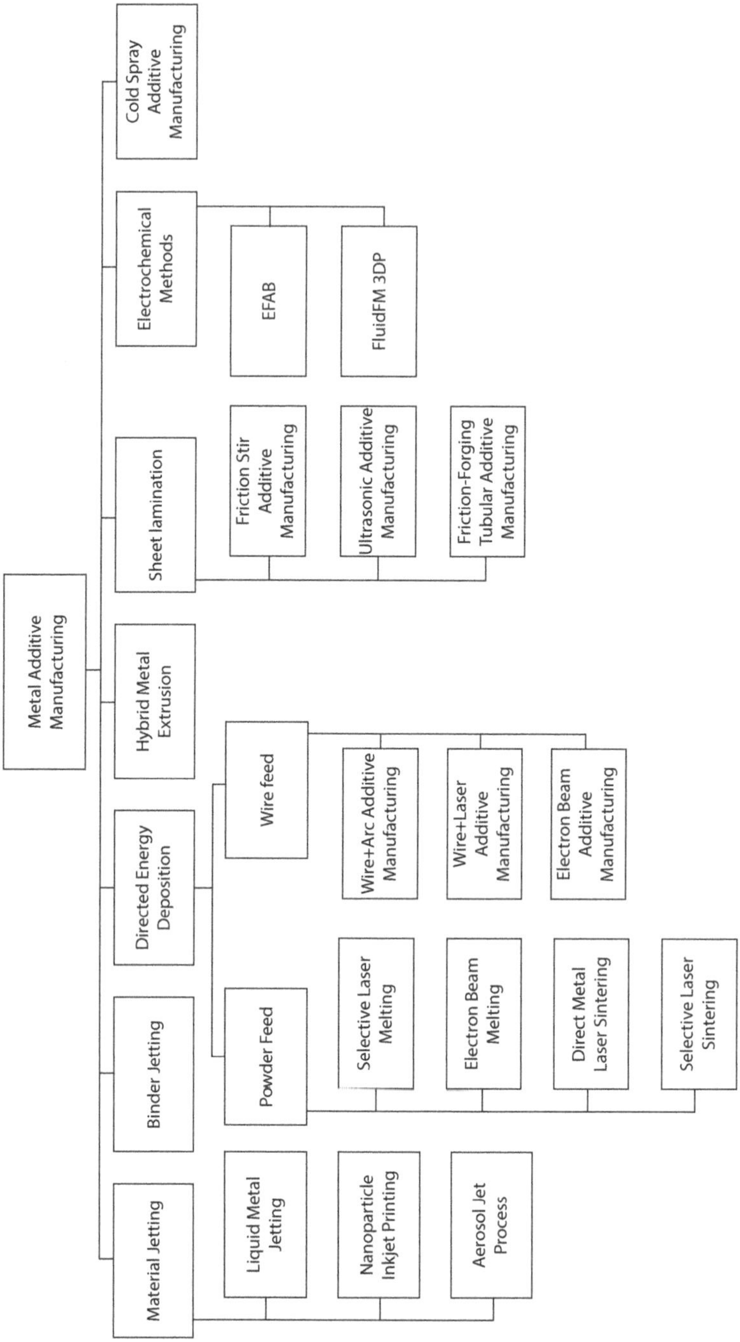

FIGURE 7.1 Classification of metal-based additive manufacturing [1].

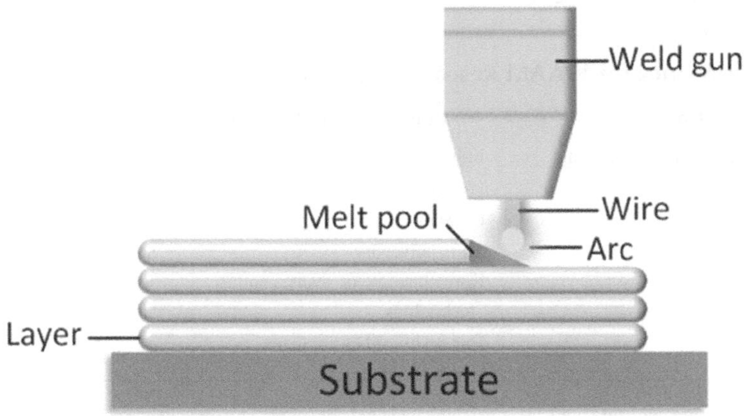

FIGURE 7.2 The mechanism of WAAM process [2].

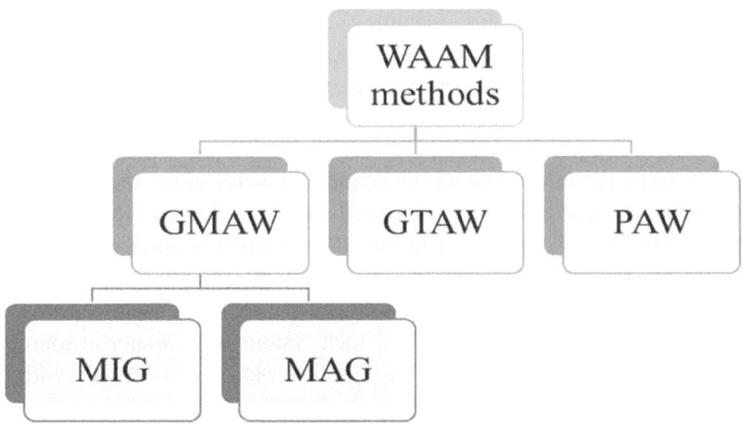

FIGURE 7.3 Types of wire arc additive manufacturing methods.

Arc welding techniques, such as gas metal arc welding (GMAW), gas tungsten arc welding (GTAW), and plasma arc welding (PAW), are utilized in WAAM processes. Depending on the kind of shielding gas employed, these techniques are further divided into metal inert gas (MIG) and metal active gas (MAG) processes. MAG uses active gases, such as carbon dioxide or oxygen mixes, whereas MIG uses inert gases, such as argon or helium. These methods are crucial for WAAM in layer-by-layer material deposition applications in additive manufacturing. Figure 7.3 represents the typical types in the WAAM process.

WAAM is a new and promising method for creating medium-to-large-scale metallic structures and components for various industries, including shipbuilding, automotive, and aerospace. Due to its cost-effectiveness, reduced material waste, and shorter lead times, it has become a viable alternative to conventional manufacturing

TABLE 7.1
Various Names for WAAM Research Institutes [3]

Alternative Name	Research Institute Name	Method Used
Hybrid layered manufacturing	Indian Institute of Technology Bombay	GMAW
WAAM	University of Wollongong	GMAW, GTAW
WAAM	Cranfield university	GMAW, GTAW, PAW
Near net shaping	Tuffs university	GMAW, PAW
3D welding	Korea Institute of science and technology	GMAW
3D welding	University of Nottingham	GMAW

techniques. Research institutions use various terms (synonyms) to refer to WAAM, which are listed in Table 7.1. It also enumerates the energy sources used.

Over the past few years, WAAM has been extensively employed to create components using various materials, including copper-based alloy wires. Along with increased manufacturing efficiency and reduced production costs, the WAAM technique offers the ability to create structures that can be extended to tens of meters and construct huge, completely dense pieces. It can also be used to maintain and repair broken parts and components. Metal wire used in the WAAM method typically costs around 10% more than the same weight of metal powder. Here, the material deposition rate of the WAAM process is contrasted with that of other AM techniques. For the WAAM process, the deposit rate is 500–10,000 g/h, whereas for SLM, EBM, and LMD processes it is 40–100, 100–300, and 150–2400 g/h, respectively [3].

Figure 7.4 depicts the schematic of arc welding processes used to additively fabricate metal components with a wire feedstock system as the material source. In the case of GMAW and PAW, the wire is fed into the weld pool coaxially with the arc, whereas in the case of GTAW, the feed wire is supplied separately.

7.3.1 MIG/MAG in WAAM

In MIG/MAG (metal inert gas/metal active gas) welding, the wire melts into the weld pool to create a bead, serving as both the electrode and the filler material. It is further referred to as WAAM, which deposits these beads in a layered, regulated mannern to produce parts with a near-net shape. Ti6Al4V can also be processed using WAAM with MIG/MAG. However, there are some difficulties, such as arc wandering with titanium. Because the active gas helps weld penetration, MAG is better suited for ferrous metals like steel, whereas MIG is usually utilized for non-ferrous metals like titanium and aluminum.

An electric arc is created between a consumable wire rod/electrode and the substrate using MIG/MAG, creating a melt pool protected by an inert gas (helium or argon) or active (mostly gas combinations such as CO_2) [5]. The addition is encouraged by melting and combining both components in the pool.

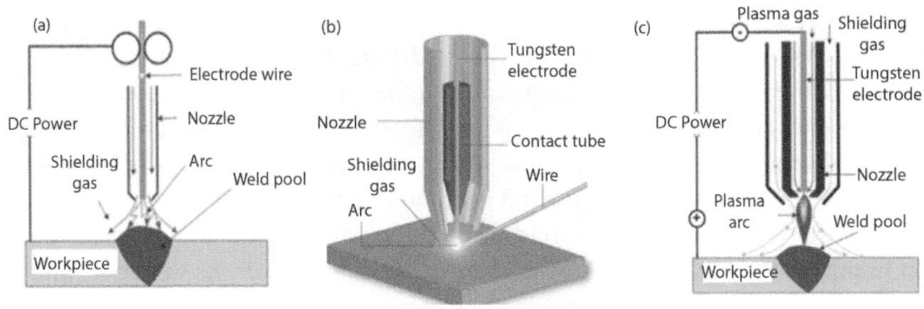

FIGURE 7.4 Various arc-based welding processes: (a) GMAW, (b) GTAW, and (c) PAW [4].

7.3.2 CMT IN WAAM

Due to its minimal heat input, cold metal transfer (CMT), a cutting-edge technology derived from the traditional MIG process, has completely changed the joining of comparable and dissimilar metals. CMT arc welding, developed by Fronius International GmbH in the early 2000s, represents a breakthrough in welding knowledge in the gas metal arc welding field, also called MIG/MAG welding. The background of CMT welding begins with GMAW, which was developed in the 1940s and used a shielding gas and a consumable electrode wire to produce an arc [6]. However, previous techniques were difficult to control and frequently produced spatter, particularly when working with thin materials. As a result, in the 1980s and 1990s, pulse GMAW techniques were developed, which pulsed the welding current between high and low levels to provide better control and enhancement of the deposited beads [7].

CMT welding is one such technology that has emerged to yield economical and fruitful results, resulting in fewer defects during component fabrication. Other advantages of the CMT process are low heat input, very little spatter, and smooth operation that can be ensured [8]. Studying various approaches to achieve a controlled transient process is essential for reducing heat input. Thus, one can achieve well-shaped and properly controlled fabrication of welded components in arc welding.

The deposition of several beads thus optimizes the process parameters depending on the user's requirement to fabricate a component properly and ideally by additive manufacturing. Using a CMT welding source and a Taguchi L9 orthogonal array of experiments, bead-on-substrate trials and identifying appropriate process parameters for effective bead shape were carried out for the AZ31 magnesium alloy [9]. Regression equations and analysis of variance (ANOVA) were used to examine the findings. The R^2 and Adj R^2 values for every response were above 90%, indicating the model's high fitness. Thus, the CMT process can produce AM components with good surface finish, dimensional accuracy, and defect-free and spatter-free medium and large sizes.

Nevertheless, the optimization of operating parameters is crucial for the large-scale production of steel alloys, aluminum alloys, titanium alloys, magnesium alloys, and other components, which are extremely valuable for sectors such as aerospace and automotive, and is possible using WAAM. The microstructure, mechanical characteristics, and corrosion performance of the magnesium alloy AZ31 made by the CMT-WAAM technique are investigated [10]. According to a microstructural study, welding conditions and cooling rates impact the AZ31's equiaxed grain structure and intergranular precipitates. With a focus on the connection between wire feed, weld speed, and bead properties, the study also examines how process parameters influence the microstructure and geometry of the beads.

The wire retraction mechanism is an essential part of cold metal transfer welding that improves the process by allowing for precise control over the weld pool and minimizing spatter. It provides a concise description of how it operates within the framework of CMT. In Fronius CMT welding, wire sense technology synchronizes wire retraction timing with the wire, acting as a sensor for position or contact changes for optimal gaps or misalignments. Springless retraction with precise motorized wire feeders occurs in milliseconds, retracting the wire fractionally to separate the droplet. It requires high-torque motors and low-latency control to accommodate high-speed, repeatable cycles. Figure 7.5 depicts the to-and-fro motion of the electrode (feeder wire) that results in supplying a low heat input to the weld pool, which is an excellent feature of the CMT technique. The retraction mechanism, which modifies the weld pool in both directions, is accurately represented by red arrow marks, even in the case of a millisecond time gap. Thus, the to-and-fro mechanism plays a key role in this method.

Due to the above-mentioned advantages, the CMT-based WAAM process extended its application across various fields. Some of the key applications are summarized in Table 7.2

Similarly, the CMT welding process has been spread through multi-wire and hybrid types of WAAM processes. The summarized analysis was made in the following description of this chapter.

7.3.3 Multi-wire in WAAM

Traditionally, WAAM provides an electric arc to melt a single wire using a single wire feeder, producing components whose chemical composition is dictated by the wire material. After specific layers are deposited, wire modifications are required when fabricating multi-material components. Increased cycle times, the introduction of surface imperfections, and numerous pauses in the deposition process are the results of this approach [12]. The multi-wire feeding WAAM approach significantly enhances WAAM's capability for producing multi-material components, thereby overcoming these challenges.

High-performance multi-material components are in high demand in the industry due to their enhanced functionality in cutting-edge applications. A single wire with a specific chemical composition is typically the only wire used in traditional WAAM, which imparts fixed qualities to the manufactured component. Due to its cost-effectiveness, flexibility, and ability to produce property differences in specific locations, multi-wire arc additive manufacturing has become a viable alternative [13]. By synthesizing knowledge from the body of existing literature, it is essential to emphasize the recent developments, research directions, and revolutionary potential

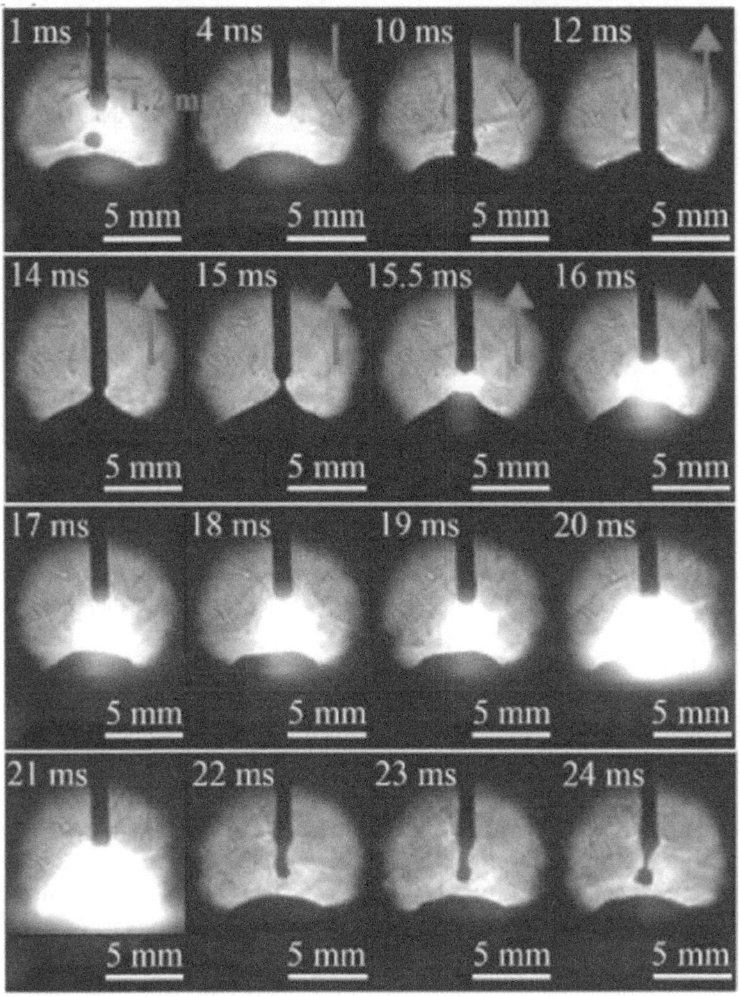

FIGURE 7.5 Retraction mechanisms in CMT welding [11].

of MWAAM in materials science and engineering. Thus, this MWAAM concept is further explored in detail. The materials and uses of MWAAM in the production of various materials, functionally gradient materials (FGMs), high entropy alloys (HEAs), shape memory alloys (SMAs), intermetallic compounds (IMCs), in-situ alloys, and other new applications are covered in this chapter. Additionally, it highlights developments in wire feeding mechanisms, deposition processes, and hybrid manufacturing approaches as it examines improvements in MWAAM procedures.

A class of engineered multi-material components, known as bimetallic structures (BMS), comprises two different metallic systems that have been carefully combined to leverage their complementary properties. These qualities include high oxidation

TABLE 7.2
Applications of CMT Welding Processes [6]

Material	Reason for Using CMT	Application
Aluminum	Precise heat control controls distortion and burn-through	Automotive component
Stainless steel	Low heat input causes low distortion	Aerospace
	Lowers the spatter for improved cleanliness	Food processing equipment
	Unique in welding thin materials	Architectural structures
Carbon steel	Can weld dissimilar metals	Structural welding
	Suitable for welding in all positions	pipelines
Titanium	Reduced post-weld cleanup due to minimal spatter	Medical devices and bio-implantation

and corrosion resistance, and thermophysical, mechanical, electrical, and optical properties [14]. Much research has been conducted on various material combinations, wire sizes, heat sources, and feedstock providers to tailor these structures for specific uses.

By enabling the creation of complex, multi-metal structures with regulated composition gradients, additive manufacturing (AM) techniques, particularly wire arc additive manufacturing (WAAM), have helped reduce the sudden interfacial flaws frequently associated with traditional fusion welding. In contrast to conventional welding techniques, MWAAM provides accurate control over dilution and microstructure development, lowering residual stresses and minimizing the creation of intermetallics. In high-performance engineering fields such as nuclear and aerospace, where exceptional corrosion resistance and mechanical dependability are crucial, the use of hybrid material systems has grown significantly. The requirement for site-specific material qualities to survive a range of operating circumstances while also lowering reliance on pricey advanced alloys drives this paradigm change, resulting in significant resource and economic efficiencies [15].

A bimetallic structure (BMS) deposits two different materials, usually in a sequential layer. Sharp transitions are frequently caused by the extreme contrast in physical and chemical characteristics at the interface, such as thermal conductivity and thermal expansion coefficients. These sudden contacts can potentially create significant residual stresses, which might encourage the formation of cracks and jeopardize the structure's mechanical stability.

The mechanical and metallurgical properties of these BMS material systems have been the subject of numerous studies. Aluminum, stainless steel, titanium, magnesium, copper alloys, and nickel-based superalloys are frequently used metal pairings widely used in the automotive, marine, and aerospace industries [16]. It has been demonstrated that introducing a slow and continuous compositional change between the materials can successfully reduce residual stresses and improve structural reliability to overcome these difficulties. FGMs have shown significant efficacy in this area since they allow for a seamless change in characteristics by progressively altering their chemical composition and microstructure across the substance's interface.

The wire arc additive manufacturing (WAAM) technique allows functionally graded materials (FGMs) to be fabricated layer by layer, varying the wire feed speed of various welding wires [17]. This work aimed to use dual-wire arc additive manufacturing (D-WAAM) to create SS 316L/Inconel 625 FGMs with different transition proportions.

7.3.4 Hybrid GMAW in WAAM

Combining the pulsed GMAW technique with the CO_2 laser creates hybrid welding (HWAAM). Arc voltage, arc current, and pulse frequency have all been adjusted so that one droplet transfers with each pulse. It is a cutting-edge welding method with extensive applications in the pipeline, telescopic, shipbuilding, and automotive industries [18]. Instead of considering either of the sources, the laser–arc hybrid process, which combines both laser and arc welding techniques, enhances the overall quality of the weld. Li et al. found that combining laser and gas metal arc welding increased the material's strength and ductility [19]. When welding austenitic stainless steel, Rayes et al. noted the significance of laser power on the metal transfer during laser–arc hybrid welding and concluded that consistent laser and arc power offered greater penetration [20].

Because a steady molten pool forms during hybrid welding, the critical welding speed is much higher than with the GMAW process. During the arc's base period, the interaction between the arc and the laser plume is seen [21]. Because of the laser plume's excellent electrical conductivity, the arc is drawn to it. The interaction impacts scatter generation because it alters the direction of the electromagnetic force. Figure 7.6 depicts the basic setup and the interaction of the laser plasma and pulsed arc.

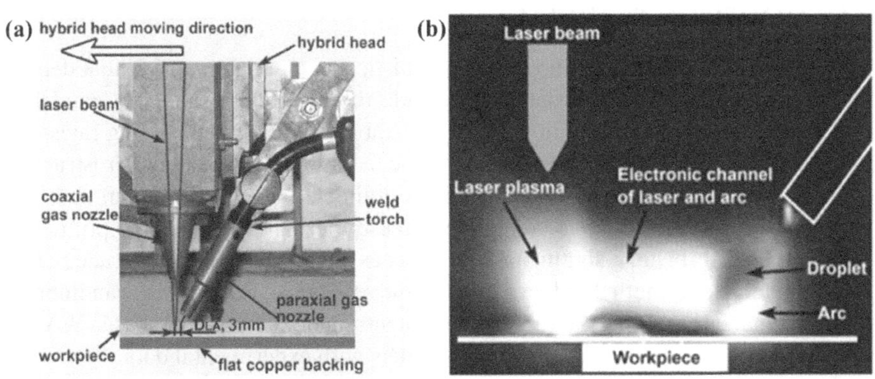

FIGURE 7.6 Hybrid welding: (a) simple setup [20] and (b) interaction of laser and arc [20].

In sheet metal welds, the size and form of a weld bead, which consists of the inner fusion boundary and the outer weld surface, are crucial strength and quality characteristics. When combined with other governing factors, the asymmetry makes it difficult to calculate the bead shape in laser–arc hybrid fillet-lap junctions using lower-order nonlinear analysis mathematical functions [18]. The parametric research clarifies the impact of input parameters on the geometry of the weld beads. The neural network demonstrates its capacity to capture the physics of the process by analyzing the weld dilution derived from the simulation data. While laser power has a negligible effect on the bead shape, welding and wire feed speeds have a significant influence. The laser enhances weld dilution, even though it uses less power, by preheating the base plate and stabilizing the welding arc.

Due to the unmatched advantages of conventional processing methods, additive manufacturing technology has advanced rapidly and continues to find new applications. However, as industrial standards rise, the manufacturing sector's urgent need for high precision, high performance, and high efficiency has become one of the main obstacles impeding the advancement of technology [22]. These demands cannot be addressed solely by laser additive manufacturing techniques. Additionally, this chapter thoroughly examines metal hybrid laser additive manufacturing technology, covering a range of topics, including multi-process hybrid laser additive manufacturing.

Hybrid wire arc additive manufacturing has rapidly transformed the contemporary metal production sector by producing medium- to large-scale metallic components. Computer-aided process planning (CAPP) is the fundamental stage in H-WAAM that contributes to increasing the output and quality of the manufactured parts [23]. Due to the higher deposition rate and the additive role of the subtractive process, the CAPP of H-WAAM must be handled differently from that of conventional additive manufacturing (AM). The conventional CAPP solutions for AM only work with files .STL format and offer limited access to process parameters. Nonetheless, a CAD/CAM system specifically designed for the SM may be utilized for the CAPP of HWAAM, provided it meets certain requirements.

7.4 MODELING IN WAAM

GMAW-based WAAM technologies are sophisticated welding methods that depend on exact process parameter control to provide the intended results. The modeling of thermal cycles, which aids in forecasting heat distribution and cooling rates during deposition, is a crucial component of these techniques. Since the microstructure of the weld or additive layer directly affects grain size, phase formation, and fault occurrence, it is imperative to comprehend these thermal dynamics to optimize it. The temperature history significantly influences other mechanical characteristics, including tensile strength, hardness, and fatigue resistance. Researchers can improve the performance and dependability of components made with GMAW-based WAAM procedures by combining computational models with experimental data.

The derivation and application of the heat transfer equation, which aims to move heat sources in arc welding, provide the basis for the analytical modeling of WAAM.

In WAAM, the Rosenthal equation is used to simulate the melt pool's geometry and heat cycle [24]. Additionally, Goldak has also conducted heat source modeling, which yields very accurate results. Of course, these models operate based on the fundamental equations that regulate the conservation of mass, momentum, and energy, which must be coupled with the actual metal part size. An example of a typical finite element environment for modeling WAAM at the macro-scale level is shown in Figure 7.7. The initial step, or pre-processing, is providing all the information (input data) required to run the program.

To optimize process parameters and part quality, GMAW-based WAAM modeling necessitates a multi-scale approach. Thermal models forecast heat flow and cooling rates to reduce residual stresses and distortion. Simulations of molten pool dynamics guarantee accurate deposition control and bead geometry. Grain growth and phase changes are linked to heat history by microstructure evolution models, and warping and cracking are avoided by thermomechanical analysis. Machine learning and data-driven algorithms optimize process parameters for improved mechanical qualities. These modeling techniques work together by connecting process, structure, and property links to create high-performance, defect-free WAAM components. Figure 7.8 summarizes these aspects.

Figures 7.9 and 7.10 illustrate the temperature variation trend at the midpoint and the residual stress distribution with the WAAM-processed thin wall SS2209 component, respectively.

The process modeling, which reproduced the cooling rates and temperature distributions (and compared them with the experimental results), helped to elucidate the phenomenon underlying these observations.

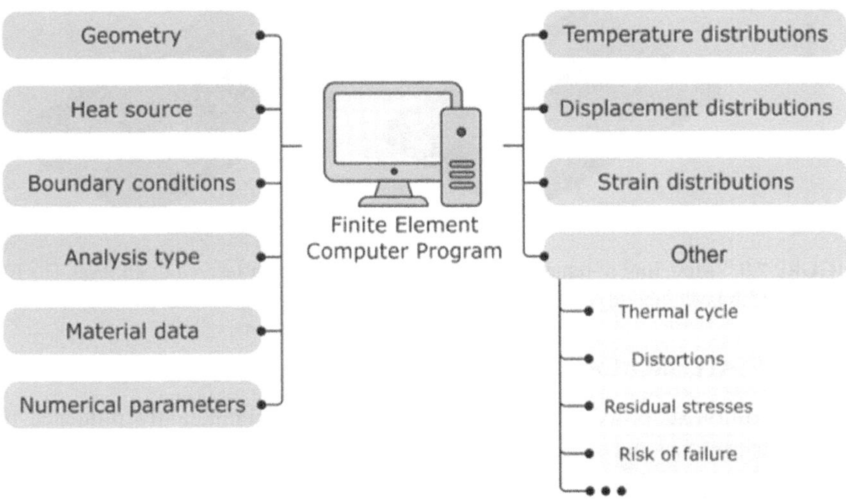

FIGURE 7.7 Modeling environment for simulating WAAM parameters [24].

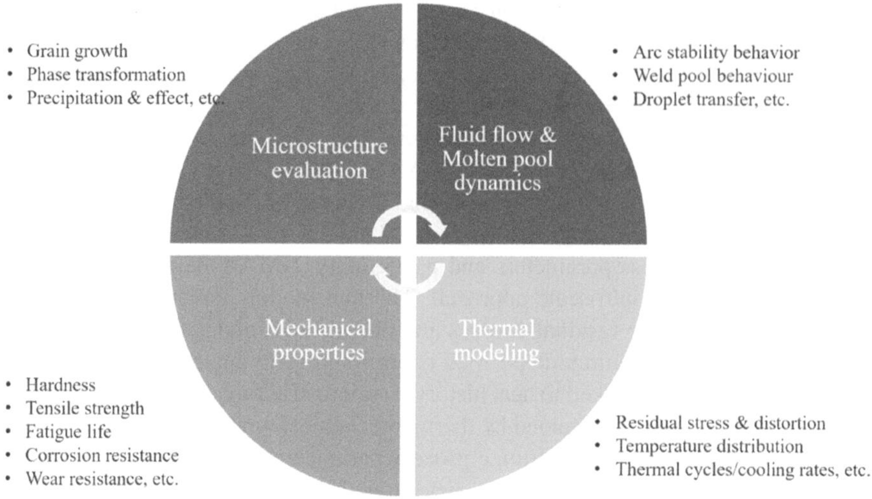

FIGURE 7.8 Various modeling aspects of GMAW.

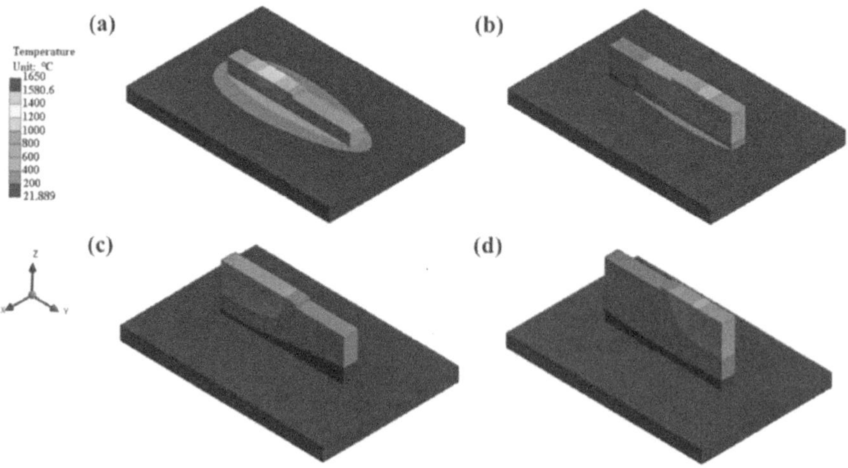

FIGURE 7.9 Modeling of temperature distribution at different layers (a) 5th layer, (b) 10th layer, (c) 15th layer, and (d) 20th layer [25].

7.5 METALLURGICAL ASPECTS

The metallurgical properties of WAAM components significantly influence the mechanical performance and structural integrity of manufactured components. The material undergoes numerous temperature cycles due to the layer-by-layer deposition process, resulting in a complex microstructural evolution. Epitaxial solidification frequently causes columnar grain formation along the build direction in steels, but

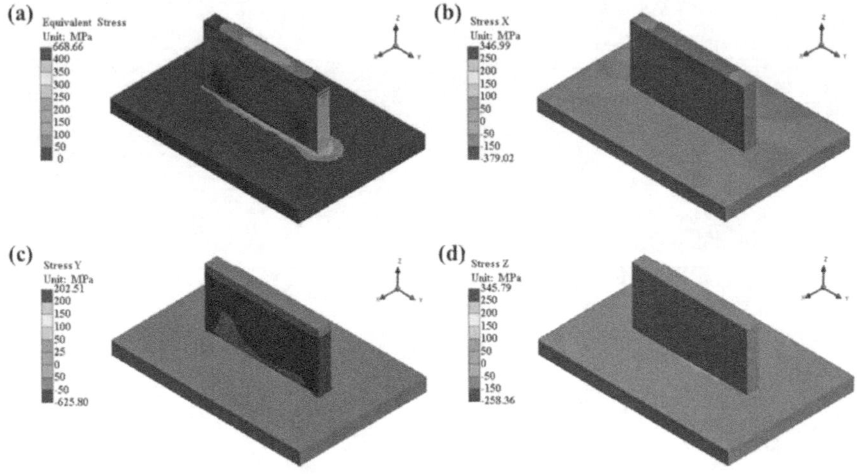

FIGURE 7.10 Modeling of residual stress distribution of WAAM component [25].

depending on cooling rates and alloy composition, equiaxed or fine-grained structures may emerge in titanium and aluminum alloys. Thermal history significantly impacts phase transformations that alter hardness and strength, such as the development of precipitate in age-hardenable alloys or the creation of martensite in high-strength steels [26].

Additionally, elemental and microsegregation are important metallurgical considerations that can lead to inhomogeneous mechanical properties, particularly in multi-pass depositions. For example, the precipitation of sigma phase or the retention of delta ferrite may impact the corrosion resistance of stainless steel. Due to solute redistribution during solidification, aluminum alloys are more susceptible to hot cracking [27]. To recover ductility and dimensional stability, post-process heat treatments such as solution annealing or stress relieving are also necessary, as residual stresses from uneven cooling can cause deformation or fracture.

Understanding the microstructural evolution of a component deposited with W-DED is accomplished by tracking both the lamination direction (also known as the build direction) and the growth. Goka et al. examine the microstructure variation and build direction for weld-deposition thin-walled components constructed of low-carbon steel [28]. This study aims to address the issue of anisotropy due to distinct thermal histories of each layer within the broader context of metal additive manufacturing. The development of a broad hypothesis for the remelt-re-solidification event, supported by thermal modeling of the process and experimental data, makes this instance unique. Repeating temperature increases and decreases in each layer deposition across the layers support this theory. In the broader context of metal additive manufacturing, this study aims to address the issue of anisotropy resulting from the unique thermal history of each layer.

The overall mechanical characteristics of FGMs were improved by implementing an optimization technique for transition gradients that excluded component regions

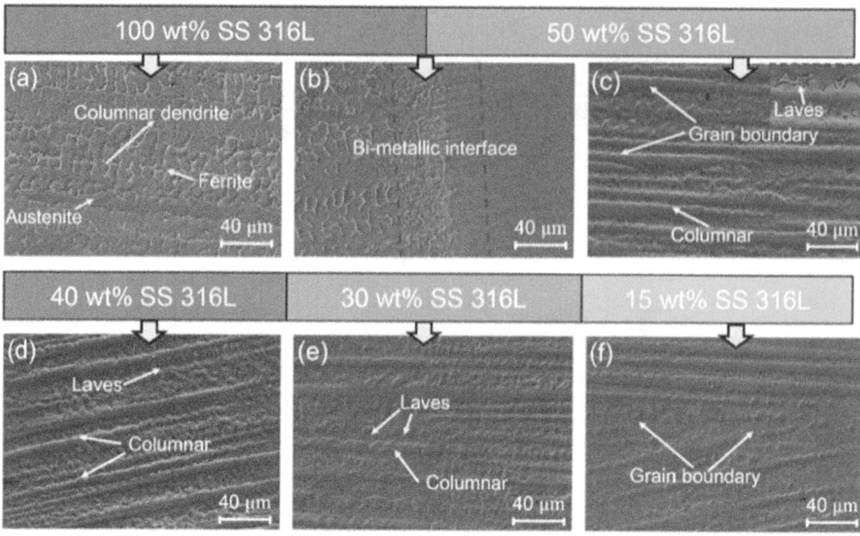

FIGURE 7.11 SEM images of AM-processed SS316L and bimetallic materials [17].

prone to defect formation (particularly cracking) while retaining other component regions [17]. Figure 7.11 represents the SEM images of AM-processed SS316L and bimetallic materials. The grain boundary cracking and the lowest microhardness and tensile characteristics were found in a 20 wt.% Inconel 625 transition gradient zone, which has a detrimental effect on the overall mechanical properties of FGMs. After that, the cracks vanished, microhardness increased, and improved tensile characteristics were achieved as the amount of Inconel 625 in the first transition area increased. The best mechanical qualities were enhanced at 50 wt.% Inconel 625 concentration. To sum up, the compositional gradient optimization approach effectively removes component areas with subpar mechanical qualities and micro defects, guaranteeing superior overall mechanical properties of FGMs.

In-situ monitoring, alloy selection, and accurate heat management are necessary to control these metallurgical characteristics. Methods like pulsed GMAW, interpass cooling management, and hybrid heat sources (such as laser-assisted WAAM) aid in microstructure refinement and defect reduction. Mechanical surface treatments and hot isostatic pressing (HIP) are post-processing techniques that improve mechanical performance and fatigue life. Optimizing GMAW–WAAM processes to create high-quality, flawless components for heavy industry, automotive, and aerospace applications requires understanding these metallurgical phenomena.

7.6 MECHANICAL PROPERTIES

In gas metal arc welding, process factors, heat cycles, and microstructural features affect the mechanical qualities of wire arc additive manufacturing components. Due to layer-by-layer deposition, WAAM components often exhibit anisotropic behavior, with mechanical properties varying depending on the build direction (vertical) and deposition direction (horizontal). While residual stresses and flaws, such as porosity or lack of fusion, can affect ductility, coarse-grain architectures typically result in tensile strength and yield strength that are comparable to or slightly lower than those of wrought materials. Hot isostatic pressing (HIP) and heat treatment (annealing, normalizing) are post-process treatments that can recover ductility and homogenize characteristics.

Hardness is highly impacted by alloy composition and cooling rates. Rapid cooling causes hard martensite to develop in high-strength steels. Precipitate coarsening causes aluminum alloys to soften in heat-affected zones (HAZs). Surface roughness, porosity, and residual stresses can frequently decrease fatigue resistance; however, methods such as machining or peening (shot or laser) can increase it by creating compressive stresses. Microstructural heterogeneity may result in lower fracture toughness and impact resistance compared to forged components; however, by refining the microstructure and minimizing defects, sophisticated hybrid techniques (such as laser-assisted WAAM and CMT) and optimized interpass temperature management can improve performance. For industrial adoption in defense, automotive, and aerospace applications, it is essential to comprehend these mechanical behaviors.

GMAW-based WAAM components exhibit anisotropic behavior in tensile testing because of the layer-by-layer deposition technique. Due to the effects of inter-pass bonding and grain orientation, the yield strength (YS) and ultimate tensile strength (UTS) in the horizontal deposition direction are typically 5–15% higher than in the vertical build direction. However, due to microporosity, inclusions, and residual stresses, ductility (% elongation) is frequently lower than that of wrought materials. Table 7.3 reveals the range of numerical representations for various materials subject to heat treatments and other property-enhancing processes as mentioned earlier.

WAAM has enormous potential for producing large-scale magnesium (Mg) alloy parts in high demand in industries such as the automotive and aeronautical fields. Goka et al. discuss the mechanical properties and corrosion behavior of CMT-WAAM-produced AZ31 magnesium alloy [10]. Microstructural investigation reveals that the WAAM-built AZ31 exhibits an intergranular precipitate-containing

TABLE 7.3
Mechanical Properties of Various Processed Materials

Material Type	Ultimate Tensile Strength	Hardness Values
Steels (LCS, SS)	450–550 MPa	180–250 HV (as deposited) 250–300 HV (cold worked, shot peened)
Aluminum alloys (Al-Si alloys)	200–300 MPa	60–90 HV (as deposited) 90–120 HV (T6 heat treated)
Titanium alloys	900–1100 MPa	300–350 HV (as deposited) 280–400 HV (stress relieved and shot peened/HIP)

equiaxed grain structure, which is a function of the welding parameters and cooling rates. The corrosion resistance properties indicate that the bottom region of the thin wall components exhibits the best resistance to corrosion, owing to a finer equiaxed grain structure, which also yields better hardness and tensile properties compared to the other regions of the thin wall.

7.7 ADVANTAGES AND LIMITATIONS OF WAAM

WAAM, with its superior material efficiency and scalability, provides large-scale metal 3D printing quickly and economically, utilizing standard welding wires for sectors like aerospace and marine. Surface roughness, anisotropic characteristics, residual tensions, and porosity are some drawbacks, though, necessitating post-processing for accuracy. It is less appropriate for high-resolution parts than powder-based AM but perfect for sturdy, near-net-shaped components. Key constraints can be lessened by process optimization and hybrid manufacturing (e.g., WAAM + machining). Table 7.4 illustrates the overall advantages and disadvantages experienced in the WAAM process.

7.8 CONCLUSIONS

In today's automated manufacturing world of Industry 5.0, automated manufacturing is shifting toward mass customization through collaborative automation of human beings and robots [29]. Since numerous manufacturing technologies are involved in welding across different industries, every new trend must be marked for different applications. AM presents significant advantages in saving raw materials and shrinking cycle time by depositing material in layers under computer control. In conclusion, GMAW-based WAAM is especially useful for heavy industrial, marine, and aerospace applications since it provides an affordable way to produce large-scale metal components with good material efficiency. Although its deposition rate and scalability are excellent, post-processing and parameter optimization are required because of issues including anisotropic characteristics, surface roughness, and residual stresses. Although hybrid manufacturing and process control developments are

TABLE 7.4
Summarized the Advantages and Limitations of the WAAM Process

Advantages	Limitations
• High deposition rate	• Anisotropy nature
• Wide material range and compatibility	• Microstructural challenges
• Material efficiency	• Surface roughness
• Cost-effective	• Minor defects
• Hybrid manufacturing	• Limited resolution
• CNC post-processing, etc.	• Residual stress and distortion, etc.

still expanding their capabilities, it is still less appropriate for high-precision applications than powder-based AM techniques. WAAM has the potential to revolutionize large-format metal additive manufacturing, thanks to continuous advancements in automation and materials.

REFERENCES

1. M. Tebianian *et al.*, "A review of the metal additive manufacturing processes," *Materials (Basel).*, vol. 16, no. 24, 2023. https://doi.org/10.3390/ma16247514.
2. T. M. T. U. Van-Thuc Nguyen, Pham Son Minh, "WAAM technique: Process parameters affecting the mechanical," *Metals (Basel).*, pp. 1–15, 2023. https://doi.org/doi.org/10.3390/met13050873.
3. D. Ding, Z. Pan, D. Cuiuri, and H. Li, "Wire-feed additive manufacturing of metal components: Technologies, developments and future interests," Int. J. Adv. Manuf. Technol., vol. 81, no. 1–4, pp. 465–481, 2015. https://doi.org/10.1007/s00170-015-7077-3.
4. D. Jafari, T. H. J. Vaneker, and I. Gibson, "Wire and arc additive manufacturing: Opportunities and challenges to control the quality and accuracy of manufactured parts," *Mater. Des.*, vol. 202, p. 109471, 2021. https://doi.org/10.1016/j.matdes.2021.109471.
5. A. F. Moreira, K. S. B. Ribeiro, F. E. Mariani, and R. T. Coelho, "An initial investigation of Tungsten Inert Gas (TIG) torch as heat source for Additive Manufacturing (AM) process," *Procedia Manuf.*, vol. 48, pp. 671–677, 2020. https://doi.org/10.1016/j.promfg.2020.05.159.
6. P. N. Bellamkonda, M. Dwivedy, and R. Addanki, "Cold metal transfer technology - A review of recent research developments," *Results Eng.*, vol. 23, no. April, p. 102423, 2024. https://doi.org/10.1016/j.rineng.2024.102423.
7. C. G. Pickin and K. Young, "Evaluation of cold metal transfer (CMT) process for welding aluminium alloy," *Sci. Technol. Weld. Join.*, vol. 11, no. 5, pp. 583–585, Sep. 2006. https://doi.org/10.1179/174329306X120886.
8. S. Hu, H. Zhang, Z. Wang, Y. Liang, and Y. Liu, "The arc characteristics of cold metal transfer welding with AZ31 magnesium alloy wire," *J. Manuf. Process.*, vol. 24, pp. 298–306, 2016. https://doi.org/10.1016/j.jmapro.2016.10.001.
9. S. Goka, M. Mallaiah, and M. J. Davidson, "Optimisation of bead morphology for cold metal transfer wire arc additive manufacturing of AZ31 magnesium alloy wires," 2024, pp. 129–137. doi: 10.1007/978-3-031-50240-8_27.
10. S. Goka, M. Manjaiah, and M. J. Davidson, "Study on mechanical, microstructural and corrosion analysis of wire arc additive manufactured AZ31 magnesium alloy," *Met. Mater. Int.*, 2024. https://doi.org/10.1007/s12540-024-01864-w.
11. Q. Gou, Z. Zhang, L. Xu, D. Wu, T. Zhang, and H. Liu, "Heat and mass transfer behavior in CMT plus pulse arc manufacturing," *Int. J. Mech. Sci.*, vol. 281, p. 109638, Jan. 2024. https://doi.org/10.1016/j.ijmecsci.2024.109638.
12. S. B. Ainapurapu, V. A. R. Devulapalli, R. P. Theagarajan, B. K. Chigilipalli, R. K. Kottala, and M. Cheepu, "Microstructure and mechanical properties of the Bimetallic Wire Arc Additively Manufactured Structure (BAMS) of SS304L and SS308L fabricated by hybrid manufacturing process," *Trans. Indian Inst. Met.*, vol. 76, no. 2, pp. 419–426, Feb. 2023. https://doi.org/10.1007/s12666-022-02695-2.
13. R. S. Tanwar and S. Jhavar, "Multi-wire additive manufacturing: A comprehensive review on materials, microstructure, methodological advances, and applications," *Results Eng.*, vol. 26, p. 104814, Jan. 2025. https://doi.org/10.1016/j.rineng.2025.104814.

14. M. R. U. Ahsan *et al.*, "Microstructures and mechanical behaviour of the bimetallic additively-manufactured structure (BAMS) of austenitic stainless steel and Inconel 625," *J. Mater. Sci. Technol.*, vol. 74, pp. 176–188, May 2021. https://doi.org/10.1016/j.jmst.2020.10.001.
15. C. Dharmendra, S. Shakerin, G. D. J. Ram, and M. Mohammadi, "Wire-arc additive manufacturing of nickel aluminium bronze/stainless steel hybrid parts – Interfacial characterisation, prospects, and problems," *Materialia*, vol. 13, p. 100834, Sep. 2020. https://doi.org/10.1016/j.mtla.2020.100834.
16. A. Rajesh Kannan, S. Mohan Kumar, R. Pramod, N. Pravin Kumar, N. Siva Shanmugam, and Y. Palguna, "Microstructure and mechanical properties of wire arc additive manufactured bi-metallic structure," *Sci. Technol. Weld. Join.*, vol. 26, no. 1, pp. 47–57, Jan. 2021. https://doi.org/10.1080/13621718.2020.1833140.
17. Q. Zhu, X. Yu, P. Yao, Y. Yue, and G. Kang, "Study on optimisation strategy for the composition transition gradient in SS 316L/Inconel 625 functionally graded materials," *Materials (Basel).*, vol. 17, no. 12, pp. 1–15, 2024. https://doi.org/10.3390/ma17122910.
18. P. Kochar, A. Sharma, T. Suga, and M. Tanaka, "Prediction and control of asymmetric bead shape in laser-arc hybrid fillet-lap joints in sheet metal welds," *Lasers Manuf. Mater. Process.*, vol. 6, no. 1, pp. 67–84, 2019. https://doi.org/10.1007/s40516-019-0081-y.
19. R. Li, Z. Li, Y. Zhu, and L. Rong, "A comparative study of laser beam welding and laser–MIG hybrid welding of Ti–Al–Zr–Fe titanium alloy," *Mater. Sci. Eng. A*, vol. 528, no. 3, pp. 1138–1142, Jan. 2011. https://doi.org/10.1016/j.msea.2010.09.084.
20. M. Gao, S. Mei, Z. Wang, X. Li, and X. Zeng, "Process and joint characterisations of laser–MIG hybrid welding of AZ31 magnesium alloy," *J. Mater. Process. Technol.*, vol. 212, no. 6, pp. 1338–1346, Jun. 2012. https://doi.org/10.1016/j.jmatprotec.2012.01.011.
21. T. Sugino, S. Tsukamoto, T. Nakamura, and G. Arakane, "Fundamental study on welding phenomena in pulsed laser-gma hybrid welding," International Congress on Applications of Lasers & Electro-Optics, Laser Institute of America, 2005, p. 302. https://doi.org/10.2351/1.5060483.
22. W. Yue, Y. Zhang, Z. Zheng, and Y. Lai, "Hybrid laser additive manufacturing of metals: A review," *Coatings*, vol. 14, no. 3, p. 315, Mar. 2024. https://doi.org/10.3390/coatings14030315.
23. R. Sarma, S. Kapil, and S. N. Joshi, "Development of a framework for computer aided design and manufacturing of 3 axis hybrid wire arc additive manufacturing," *Mater. Today Proc.*, vol. 62, no. P14, pp. 7625–7634, 2022. https://doi.org/10.1016/j.matpr.2022.05.011.
24. R. F. V. Sampaio, J. P. M. Pragana, I. M. F. Bragança, C. M. A. Silva, C. V. Nielsen, and P. A. F. Martins, "Modelling of wire-arc additive manufacturing – A review," *Adv. Ind. Manuf. Eng.*, vol. 6, no. May, 2023. https://doi.org/10.1016/j.aime.2023.100121.
25. H. Zhang, K. Peng, W. Wu, L. Yu, R. A. Singh, and X. Chen, "Efficient additive manufacturing of 2209 duplex stainless steel using multi-stranded wire plasma arc: Numerical simulation, microstructure and mechanical properties," *Vacuum*, vol. 238, Nov. 2024, 2025. https://doi.org/10.1016/j.vacuum.2025.114251.
26. L. Wang, J. Xue, and Q. Wang, "Correlation between arc mode, microstructure, and mechanical properties during wire arc additive manufacturing of 316L stainless steel," *Mater. Sci. Eng. A*, vol. 751, pp. 183–190, Mar. 2019. https://doi.org/10.1016/j.msea.2019.02.078.
27. G. L. Knapp, M. Gussev, A. Shyam, T. Feldhausen, and A. Plotkowski, "Microstructure, deformation and fracture mechanisms in Al-4043 alloy produced by laser hot-wire additive manufacturing," *Addit. Manuf.*, vol. 59, p. 103150, 2022. https://doi.org/10.1016/j.addma.2022.103150.

28. J. D. Kulkarni, S. B. Goka, P. K. Parchuri, H. Yamamoto, K. Ito, and S. Simhambhatla, "Microstructure evolution along build direction for thin-wall components fabricated with wire-direct energy deposition," *Rapid Prototyp. J.*, vol. 27, no. 7, pp. 1289–1301, 2021. https://doi.org/10.1108/RPJ-04-2020-0085.
29. S. Goka, S. Q. Moinuddin, M. Cheepu, and A. K. Dewangan, "Welding practices in industry 5.0: Opportunities, challenges, and applications," in *Automation in Welding Industry*. Wiley, 2024, pp. 263–279. doi: 10.1002/9781394172948.ch15.

8 Challenges and Opportunities in GMAW

8.1 BACKGROUND

Welding is one of the oldest techniques used to join to metals. There are different kinds of welds, and gas metal arc welding or GMAW is one of them. "Gas metal arc welding (GMAW), also referred to as MIG (metal inert gas), or wire feed is a process where a consumable electrode (wire) is automatically fed through a welding gun along with a continuous flow of a shielding gas" [1]. GMAW like any other technique, has its set of challenges, such as defects like porosity and cracking caused by uncontrolled parameters, process monitoring, the difficulty of getting the welding gun into tight spaces, and having to weld in breezy conditions (without the shield). While there are challenges with this method, there are opportunities as well that can improve some of the shortcomings of GMAW. For example, the use of AI and ML to monitor the process parameters such as voltage and current is an opportunity that can reduce the need for manual process monitoring. "Most of the applications developed using machine learning for process monitoring can identify anomalous process performance, which may indicate weld defects [2]." Another opportunity is reducing the size of the GMAW welding equipment (miniaturization) for welding in more intricate spaces. Current GMAW machines are not portable, and using them in constricted places can be demanding for the operators. This chapter aims to elaborate on the various challenges and opportunities within the GMAW realm, such as

1. Miniaturization of GMAW equipment
2. Welding in space
3. Quality control
4. Applications of ML
5. Future scope

8.2 MINIATURIZATION OF GMAW EQUIPMENT

Before we discuss the miniaturization of GMAW, let us understand why this is required in the first place. Let us discuss the need, importance, advantages, and applications of miniaturization of GMAW welding machines.

Gas metal arc welding (GMAW), also known as metal inert gas (MIG) welding, is a popular welding process that uses an electric arc to join metals. It is favored for its versatility, ease of use, and ability to produce strong, high-quality welds. Traditional

DOI: 10.1201/9781003778844-8

GMAW welding machines are often large and heavy, making them less suitable for work in confined spaces or in environments where mobility is essential. This has led to the trend of miniaturizing GMAW welding machines, which involves reducing their size and weight while maintaining their performance. In this illustration, we will explore the need, importance, advantages, and applications of miniaturized GMAW welding machines.

8.2.1 The Need for Miniaturization

8.2.1.1 Mobility and Portability

The demand for welding services across various industries often requires welders to move from one location to another. In industries such as construction, repair, and shipbuilding, welding tasks must be performed in multiple locations, often in confined or hard-to-reach areas. The need for portable GMAW welding machines is critical in these situations, as moving traditional, heavy equipment is labor-intensive and time-consuming. Miniaturized GMAW machines offer the advantage of portability, making it easier for welders to transport their equipment between job sites.

8.2.1.2 Space Constraints and Ergonomics

Many work environments, such as automotive repair shops, small workshops, and industrial plants, have limited space for large equipment. In these settings, a compact GMAW machine is necessary to ensure efficient use of space. Miniaturized machines allow for the same high-quality welding in smaller footprints, making it possible to set up a welding station without occupying excessive space. Reducing the weight of the welding equipment such as the torch has a significant impact on the ergonomics of the welders [3].

8.2.1.3 Energy Efficiency

Large GMAW machines often consume significant amounts of energy, which can be a concern in industries aiming to reduce their carbon footprint or lower operational costs. Miniaturized machines are designed with energy efficiency in mind, offering reduced power consumption while maintaining the necessary welding output. This is important for both economic and environmental reasons, as industries look to minimize energy use and increase sustainability.

8.2.2 The Importance of Miniaturization

8.2.2.1 Increased Accessibility

The miniaturization of GMAW welding machines makes the technology accessible to a broader range of users, including small businesses, hobbyists, and educational institutions. Smaller, more affordable machines enable businesses and individuals who may not have the budget or space for traditional welding equipment to perform GMAW welding tasks. This expands the use of the technology to areas such as DIY repairs, home workshops, and schools, where full-sized machines may be impractical.

8.2.2.2 Enhanced Usability in Remote Locations

Remote locations, such as construction sites, pipelines, or offshore platforms, often require portable equipment due to limited access to resources. Miniaturized GMAW welding machines allow for easy transportation to these locations, ensuring that essential welding tasks can be performed without the need for cumbersome, stationary equipment. Their compact size is also advantageous in remote or rural areas where space is constrained, and power availability may be limited.

8.2.2.3 Cost-Effectiveness

By reducing the size and weight of GMAW welding machines, manufacturers can often lower production costs, leading to more affordable products for end users. This cost-effectiveness is particularly important for small- and medium-sized businesses that need to manage budgets while maintaining the ability to perform high-quality welding work. Miniaturized machines typically have lower maintenance costs and require fewer resources, further reducing the financial burden on users.

8.2.3 Advantages of Miniaturization

8.2.3.1 Enhanced Portability

One of the most significant advantages of miniaturized GMAW machines is their portability [4]. These compact machines are much lighter than their traditional counterparts, allowing welders to move them easily across job sites or between workstations. This is particularly beneficial in industries like automotive repair, construction, and shipbuilding, where mobility is essential.

8.2.3.2 Space Efficiency

Miniaturized GMAW machines are ideal for small workshops, garages, or confined industrial environments where space is at a premium. Their compact design allows them to be set up in tight spaces without sacrificing performance, making them perfect for environments where large equipment would be cumbersome or impractical.

8.2.3.3 Lower Power Consumption

Smaller GMAW machines typically consume less energy than larger ones. This leads to cost savings on electricity bills, making them more energy efficient. In industries that prioritize sustainability or have limited power resources, such as remote construction sites or eco-conscious businesses, miniaturized machines offer a greener alternative.

8.2.3.4 Precision and Control

Despite their smaller size, miniaturized GMAW machines are often equipped with advanced features such as digital controls, preset modes, and improved arc stability.

These features give welders greater control over the welding process, leading to more precise and consistent results. This is particularly useful in applications where accuracy is critical, such as aerospace or automotive manufacturing.

8.2.3.5 Reduced Operator Fatigue

The lighter weight and more compact design of miniaturized GMAW machines help reduce operator fatigue. Welders can move the machine more easily and position it closer to their work without straining themselves, leading to better ergonomics and safer working conditions. This is particularly beneficial for welders who perform long-duration tasks.

8.2.4 Applications of Miniaturized GMAW Welding Machines

8.2.4.1 Automotive Industry

In the automotive industry, GMAW welding is frequently used to repair car frames, body panels, and exhaust systems. Miniaturized machines offer increased mobility, allowing welders to move around the vehicle easily and reach tight spots, such as engine compartments. This makes them ideal for automotive repair shops and assembly lines, where both space and time are limited.

8.2.4.2 Construction and Fabrication

Construction and fabrication industries benefit significantly from the portability of miniaturized GMAW machines. Whether it's welding steel beams, railings, or structural components, smaller machines can be easily transported to different areas of a construction site. This improves productivity and reduces downtime, as workers can carry the machine where it is needed without the logistical challenges of moving large, stationary equipment.

8.2.4.3 Shipbuilding and Offshore Applications

"The shipbuilding and marine industry play a crucial role in global trade and transportation, requiring the construction of robust and reliable vessels and marine structures" [5]. Shipbuilding and offshore platforms require welding work in confined spaces and often in remote or difficult-to-access areas. Miniaturized GMAW machines are well-suited to these environments because of their portability and ability to work in tight areas. The machines' precision and control also make them ideal for handling complex welding tasks on materials like steel and aluminum.

8.2.4.4 Maintenance and Repair Services

Miniaturized GMAW machines are highly valuable in maintenance and repair services, particularly in industrial settings where heavy machinery or infrastructure needs regular welding repairs. Technicians can easily transport these machines to various locations, reducing downtime and allowing for rapid repairs. The portability of these machines is essential for companies that need to keep their operations running smoothly with minimal disruption.

8.2.4.5 DIY and Hobbyist Projects

For hobbyists, artists, and small-scale manufacturers, miniaturized GMAW machines offer an affordable and practical option for metalworking projects. Their compact size and ease of use make them ideal for home workshops, DIY repairs, and artistic metalwork, where large machines would be overkill.

8.2.4.6 Conclusion

The miniaturization of GMAW welding machines has revolutionized the welding industry by offering compact, energy-efficient, and portable alternatives to traditional machines. The need for miniaturization stems from the growing demand for mobility, space efficiency, and cost savings across a wide range of industries. The advantages of these smaller machines, such as enhanced portability, precision, and lower power consumption, make them invaluable in various applications, from automotive repair and construction to shipbuilding and small-scale manufacturing. As industries continue to evolve, the trend toward miniaturization will likely play an increasingly important role in the future of welding technology.

8.3 GMAW IN SPACE: POTENTIAL, ADVANTAGES, AND CHALLENGES

The possibility of using GMAW in space is an emerging field of research that has gained attention due to humanity's growing interest in space exploration, satellite construction, and even long-term colonization of other celestial bodies. GMAW could play a vital role in these missions by enabling the construction and repair of space structures while in orbit or on extraterrestrial surfaces. This study will explore how GMAW can be used in space, why it is a good idea, the advantages it offers, its challenges, and the future research required to overcome these challenges.

8.3.1 How GMAW Can Be Used in Space

8.3.1.1 Construction of Space Habitats and Structures

As space missions become longer and more complex, there will be a growing need for in-situ construction and repair of habitats, spacecraft, and space stations. GMAW could be used to weld metal components together in space, allowing for the assembly of large structures like space stations, habitats, or even satellites while in orbit. GMAW would allow astronauts to perform tasks such as repairing damage to spacecraft, fabricating parts on demand, or building large solar panels and antennas for space stations [6].

8.3.1.2 On-Demand Repairs and Maintenance

Spacecraft and satellites are exposed to a range of environmental hazards, including micrometeoroids, space debris, and radiation, all of which can cause structural damage. GMAW could be used to perform emergency repairs on damaged spacecraft or space station modules without the need to return to Earth. The ability to weld components in space would extend the lifespan of missions by allowing astronauts to fix critical components without requiring Earth-based resources.

8.3.1.3 Lunar and Martian Construction

With the possibility of human missions to the Moon and Mars, GMAW could be utilized to build infrastructure on extraterrestrial surfaces. For example, GMAW could be used to assemble metal structures, such as habitats, landing pads, or mining equipment, using materials either brought from Earth or extracted from the local environment. The process could be a key part of future in-situ resource utilization (ISRU) strategies that aim to reduce reliance on Earth-based supplies by using materials available in space.

8.3.2 Why GMAW in Space Is a Good Idea

8.3.2.1 Versatility across Materials

GMAW is highly versatile and can be used to weld a wide variety of materials, including aluminum, steel, and titanium, which are commonly used in space structures. Its adaptability means that it can handle the diverse materials needed for different components of spacecraft, habitats, and other space infrastructure. The process also allows for high-quality, reliable welds that are crucial for the safety and longevity of space missions.

8.3.2.2 Automation Potential

One of the major benefits of GMAW is its potential for automation, which is especially important in space. With the development of robotic systems, GMAW could be integrated into autonomous or remotely controlled robots that perform welding tasks in space. This could reduce the need for astronauts to perform physically demanding and dangerous extravehicular activities (EVAs), minimizing their exposure to space hazards [7]. Automated GMAW systems could also be deployed for continuous construction tasks, such as assembling space stations or mining infrastructure.

8.3.2.3 Long-Term Cost Savings

By enabling in-situ repairs and construction, GMAW could reduce the need for expensive resupply missions from Earth. The ability to fabricate parts and repair damage on-site means that missions could be more self-sufficient, which is critical for deep space missions or planetary colonies where resupply from Earth may take months or even years. This would lead to long-term cost savings, making space exploration more sustainable and economically viable.

8.3.3 Advantages of Using GMAW in Space

8.3.3.1 Weight and Resource Efficiency

Transporting materials and machinery into space is extremely costly, as launch payloads are limited by the mass and volume of rockets. GMAW machines can be designed to be compact and relatively lightweight, making them ideal for space missions. In addition, GMAW allows for precise material usage, minimizing waste and maximizing the efficiency of the resources available in space.

8.3.3.2 Continuous Welding

GMAW provides continuous, high-speed welding that is efficient for large-scale construction and repair tasks. This continuous operation could be particularly beneficial in space, where long-duration tasks need to be completed quickly to minimize astronaut exposure to hazardous environments. The continuous feed of filler material also allows for consistent, high-quality welds, which are critical for the structural integrity of space installations.

8.3.3.3 Remote or Robotic Operation

Given the hostile environment of space, GMAW could be controlled remotely or by robots. This would reduce the need for astronauts to be directly involved in welding tasks, enhancing safety. Robotic systems equipped with GMAW could also work in space environments for extended periods, enabling long-term construction projects without the need for human intervention.

8.3.4 Shortcomings of Using GMAW in Space

8.3.4.1 Vacuum Environment

One of the primary challenges of using GMAW in space is the vacuum environment. GMAW requires a shielding gas to protect the weld from atmospheric contamination. In the vacuum of space, traditional shielding gases would quickly dissipate. This problem would need to be addressed by developing new welding techniques or protective mechanisms that can shield the weld in a vacuum. For example, special enclosures could be developed to create localized atmospheric conditions that allow GMAW to function.

8.3.4.2 Microgravity Effects

In space, microgravity can affect the behavior of molten metal during the welding process. Without gravity to pull the molten metal into place, it may form irregular weld pools or behave unpredictably. This could lead to weak or inconsistent welds. Research is needed to understand how molten metal behaves in microgravity and how welding parameters (such as arc voltage, wire feed speed, and current) might need to be adjusted to compensate. "However, because the different environments, such as microgravity and vacuum, change the welding phenomena, the welding technique in space has not been completely established yet" [8].

8.3.4.3 Radiation Exposure

The high levels of radiation in space present another challenge for GMAW, particularly for the electronic components of welding machines. Prolonged exposure to radiation can degrade the performance of electronic systems and even cause malfunctions. To address this, welding equipment will need to be radiation-hardened or shielded to ensure reliable operation in space environments.

8.3.5 Future Research Needs

8.3.5.1 Shielding Solutions for Vacuum Environments

Future research should focus on developing innovative ways to protect welds in the vacuum of space. This could involve the development of new materials for welding that are less susceptible to contamination or the creation of portable shielding chambers that maintain a controlled atmosphere around the weld area. These innovations would ensure that weld quality remains high even in the absence of Earth's atmosphere.

8.3.5.2 Behavior of Molten Metal in Microgravity

Further studies are needed to understand how molten metal behaves in microgravity and how welding processes should be adapted for space. This could involve experiments conducted in microgravity environments, such as on the International Space Station (ISS), to refine GMAW techniques for space use. Research should also explore the possibility of using magnetic fields or other forces to control molten metal in space.

8.3.5.3 Radiation-Resistant Electronics

Developing electronics that can withstand the harsh radiation of space will be crucial for ensuring that GMAW machines operate reliably during long-term missions. Future research should focus on creating robust, radiation-hardened systems that are capable of functioning in deep space or on planetary surfaces.

8.3.5.4 Automation and Robotic Integration

Given the risks to human astronauts, future research should explore the integration of GMAW with robotic systems. These systems could be controlled remotely from Earth or from within spacecraft, allowing for autonomous or semi-autonomous welding tasks. This research would focus on creating intelligent robotic welders capable of adjusting to varying conditions and performing complex tasks in space.

8.3.5.5 Conclusion

GMAW has the potential to be an essential tool for space exploration, allowing for the construction and repair of critical infrastructure in orbit, on the Moon, and on Mars. While there are challenges associated with the vacuum environment, microgravity, and radiation exposure, ongoing research and innovation can help overcome these hurdles. The advantages of using GMAW in space, including its versatility, potential for automation, and cost-effectiveness, make it an attractive option for future space missions. As space exploration continues to advance, GMAW will likely play a key role in enabling humanity to build a sustainable presence beyond Earth.

8.4 QUALITY DEFECTS IN GMAW: ROOT CAUSES AND SOLUTIONS

Gas metal arc welding (GMAW) is widely used for its efficiency and ability to weld a variety of metals. However, several quality defects may arise during the welding process, affecting the integrity and strength of the weld. Understanding the root causes of these defects and applying solutions are essential to ensure high-quality results.

8.4.1 Porosity

Description: Porosity refers to the presence of small gas pockets or voids within the weld metal, weakening the weld and making it prone to failure.
Root Causes:

- Contaminated base metal (oil, rust, dirt)
- Improper shielding gas coverage
- Excessive wire feed speed

Solutions:

- Clean the base metal thoroughly before welding
- Ensure proper shielding gas flow and nozzle positioning
- Adjust wire feed speed for optimal weld formation

Potential Solutions: Advanced real-time monitoring systems that detect gas flow issues and contamination could automatically adjust the welding parameters to prevent porosity [9].

8.4.2 Incomplete Fusion

Description: Incomplete fusion occurs when the weld metal fails to bond properly with the base metal or between weld passes, leading to weak joints.
Root Causes:

- Incorrect welding parameters (low heat input)
- Improper electrode angle or travel speed
- Presence of contaminants on the base material

Solutions:

- Increase voltage or amperage to ensure sufficient heat
- Adjust electrode angle and travel speed for better penetration
- Clean the base material before welding

Potential Solutions: Automated systems with machine learning capabilities could detect insufficient heat input or improper welding angles, adjusting settings in real time to achieve complete fusion.

8.4.3 Undercutting

Description: Undercutting is a groove that forms at the weld toe, reducing the weld's strength by thinning the base metal.
Root Causes:

- Excessive heat input
- High welding speed
- Improper electrode angle

Solutions:

- Reduce heat input by lowering voltage or amperage
- Decrease welding speed to allow better metal deposition [10]
- Correct electrode angle for a smoother weld bead

Potential Solutions: Development of heat-resistant filler materials or improved heat control mechanisms could help prevent excessive undercutting.

8.4.4 Spatter

Description: Spatter involves small metal droplets that are ejected from the weld and stick to the workpiece, affecting the weld's appearance and quality.
Root Causes:

- High welding current
- Poor shielding gas coverage
- Incorrect wire feed speed

Solutions:

- Optimize current and wire feed speed
- Improve shielding gas coverage by ensuring correct gas flow rates
- Use anti-spatter sprays to reduce cleanup time

Potential Solutions: New shielding gas mixtures that reduce spatter formation or more advanced torches designed to minimize spatter during high-speed welding could be researched.

8.4.5 Conclusion

Current defects in GMAW, such as porosity, incomplete fusion, undercutting, and spatter, can often be mitigated with proper techniques, equipment, and parameter adjustments. Future solutions may involve greater automation, real-time monitoring, and adaptive systems that use artificial intelligence to optimize the welding process on the fly, improving both quality and productivity [11].

8.5 LEVERAGING MACHINE LEARNING (ML) AND ARTIFICIAL INTELLIGENCE (AI) TO IMPROVE GMAW

Gas metal arc welding (GMAW) is a highly adaptable and efficient welding process widely used across industries such as automotive, aerospace, and construction. While GMAW already offers several advantages, incorporating machine learning (ML) and artificial intelligence (AI) into the process can significantly enhance efficiency, quality, cost-effectiveness, and process control. The use of AI-driven technologies can help make GMAW more foolproof, leading to fewer defects, better process optimization, and overall improved welding outcomes. This study discusses how ML and AI can revolutionize GMAW, focusing on key areas such as efficiency, quality, cost, and control.

8.5.1 Efficiency Improvements through AI-Driven Automation

8.5.1.1 Real-Time Process Monitoring and Adjustment

AI algorithms can monitor the GMAW process in real time, analyzing data such as arc voltage, current, wire feed speed, and temperature. These systems can automatically adjust welding parameters based on feedback to maintain optimal conditions. This leads to:

- Reduced Operator Intervention: AI can adjust variables like speed or heat input dynamically, making the process smoother and more consistent.
- Enhanced Productivity: With continuous monitoring, AI systems can immediately detect inefficiencies and correct them, allowing the welding process to run more efficiently and without interruptions.

8.5.1.2 Robotic Welding and Adaptive Learning

By integrating ML with robotic welding systems, machines can learn from vast datasets of previous welds to improve future performance. For example:

- AI can analyze successful welds to identify patterns and predict the optimal parameters for different materials, environments, or conditions.
- AI-powered robots can adapt to slight variations in workpieces or environmental factors, improving efficiency even in variable conditions. This

adaptability is especially useful in automated production lines where high throughput is critical.

8.5.1.3 Predictive Maintenance

AI can also track the condition of welding equipment over time and predict when maintenance is required, reducing unexpected downtime. Machine learning models can analyze data from sensors embedded in the welding machinery to predict when components like welding torches or feed mechanisms are likely to fail, ensuring they are serviced before breakdowns occur.

8.5.2 Enhancing Weld Quality with AI

8.5.2.1 Defect Detection and Prevention

AI systems equipped with computer vision can inspect welds in real time, identifying defects such as porosity, incomplete fusion, or undercutting. These systems:

- Use cameras and sensors to monitor the welding arc, weld pool, and surrounding area, flagging defects as soon as they occur.
- Automatically adjust welding parameters to prevent further defects. For instance, if the system detects a poor weld bead formation, it can adjust the wire feed speed or voltage to correct the issue.

8.5.2.2 Adaptive Process Control

ML models can learn the optimal welding parameters for specific materials and processes, ensuring high-quality welds under varying conditions. By processing historical data from thousands of welds, AI can:

- Predict the right settings for different alloys, thicknesses, or joint configurations, ensuring consistent quality across different jobs.
- Automatically adjust to environmental factors such as changes in temperature, humidity, or material surface conditions that might affect weld quality.

8.5.2.3 Improved Consistency

AI helps eliminate variability caused by human error, making the welding process more consistent. Whether in manual welding or automated systems, AI-driven controls ensure that every weld is performed with the same precision, reducing the occurrence of defects and the need for rework.

8.5.3 Cost Reduction through AI Integration

8.5.3.1 Waste Reduction

AI can optimize the use of welding materials (wire, shielding gas, energy) by maintaining ideal welding conditions and preventing overuse. Through data analysis and real-time feedback, AI can:

- Minimize spatter, resulting in reduced waste of filler material.
- Optimize gas usage by adjusting flow rates based on real-time monitoring, ensuring efficient shielding gas coverage without unnecessary overconsumption.

8.5.3.2 Reduced Labor Costs

AI-powered systems require less manual oversight, which can lead to significant labor cost savings [12]. In automated production environments:

- Fewer operators are needed to oversee multiple welding robots.
- The need for skilled welders to adjust machine settings is reduced, as AI can dynamically adjust parameters based on real-time data.

8.5.3.3 Minimizing Rework and Scrap

By ensuring more consistent weld quality and detecting defects early, AI minimizes the need for costly rework and reduces the production of scrap parts. Fewer defective parts translate into cost savings by reducing material loss and time spent on corrective actions.

8.5.4 Process Control and Stability through Machine Learning

8.5.4.1 Predictive Models for Optimal Welding Parameters

Machine learning algorithms can analyze historical welding data to build predictive models that suggest the best welding parameters for specific tasks. This makes the process more stable and predictable:

- ML models can take into account variables such as material thickness, joint configuration, and environmental factors to predict the ideal voltage, current, and wire feed speed.
- These predictive models reduce the trial-and-error phase in setting up new welding jobs, allowing operators to begin welding with the best settings from the start [13].

8.5.4.2 Closed-Loop Control Systems

Closed-loop control systems, enhanced by AI, provide feedback to the welding machine during operation, allowing for continuous adjustment of parameters such as arc length, travel speed, or current. These systems:

- Ensure the welding process remains stable even as conditions change, such as when there is variation in material properties or environmental conditions.
- Automatically correct deviations in real time, preventing small issues from escalating into major defects.

8.5.4.3 Data-Driven Process Optimization

AI can process vast amounts of data from previous welding operations, using it to optimize future welding processes. This data-driven approach enables:

- The identification of patterns that lead to successful welds helps to standardize best practices across operations.
- Continuous improvement, as the system learns from new data and refines its control strategies over time.

8.5.5 Fool Proofing the GMAW Process

8.5.5.1 Operator Error Mitigation

AI-driven GMAW systems can significantly reduce the impact of human error. By automating parameter selection and process control, AI can:

- Guide less experienced operators by providing recommendations or adjusting settings automatically, reducing the learning curve.
- Prevent errors caused by incorrect parameter settings, as the system can recognize and correct suboptimal conditions before they affect the weld.

8.5.5.2 AI-Assisted Training

Machine learning algorithms can also be used to develop AI-assisted training systems for welders. These systems:

- Provide real-time feedback to trainees, helping them improve their techniques more rapidly.
- Use simulations to teach operators how different welding parameters affect the final product, making the learning process more efficient.

8.5.5.3 Adaptive Systems for Variable Conditions

AI makes GMAW systems more robust in handling variable conditions, such as changes in material thickness, joint geometry, or even the working environment. With AI, welding systems can:

- Automatically detect and adapt to these variations, ensuring that weld quality remains high regardless of changing conditions.
- Provide alerts or recommendations if conditions exceed certain tolerances, further reducing the chance of defects.

8.5.5.4 Conclusion

The integration of Machine Learning and Artificial Intelligence into Gas Metal Arc Welding holds the potential to revolutionize the industry by making the process more efficient, improving quality, reducing costs, and enhancing process control.

AI can automate many aspects of the welding process, from real-time defect detection to adaptive control, ensuring consistent, high-quality welds even in challenging environments. By minimizing human error and optimizing resource use, AI can help make GMAW foolproof, increasing the reliability and productivity of welding operations. As AI and ML technologies continue to evolve, their application in GMAW will unlock new levels of precision, efficiency, and scalability, transforming the future of welding.

8.6 THE FUTURE OF GMAW: INNOVATIONS IN TOOLS, TECHNOLOGIES, METHODS, AND EQUIPMENT

The future of gas metal arc welding (GMAW) is poised for transformative advancements driven by the integration of cutting-edge tools, technologies, and methods. As industries demand higher precision, efficiency, and flexibility in welding processes, GMAW is set to evolve through the incorporation of advanced automation, digitalization, and intelligent systems.

8.6.1 AI and Machine Learning for Process Optimization

AI and machine learning (ML) will continue to shape the future of GMAW by enabling smarter, more adaptive welding systems. These technologies will allow for real-time monitoring and adjustment of welding parameters, ensuring optimal performance in dynamic environments. AI-driven systems will detect anomalies, predict defects, and fine-tune variables like voltage, current, and wire feed speed on the fly, ensuring consistent weld quality with minimal human intervention.

8.6.2 Robotics and Collaborative Welding

The use of robotics in GMAW is expected to become more widespread, with a particular focus on collaborative robots (cobots) that can work alongside human welders. These cobots will enhance productivity by performing repetitive, complex tasks while human operators handle high-level decision-making [14]. Robotic systems will integrate vision-based systems to guide welds with greater precision, especially in challenging geometries or hard-to-reach areas.

8.6.3 Advanced Sensor Technology

The future of GMAW will also see the use of advanced sensors that provide real-time feedback on critical parameters such as arc stability, temperature, and weld pool behavior. These sensors will be integrated into welding machines to automatically adjust settings, leading to fewer defects and higher efficiency. Technologies like infrared and laser-based sensors will enhance the ability to monitor weld integrity during and after the process.

8.6.4 Augmented Reality (AR) and Virtual Reality (VR) for Training

Augmented and virtual reality tools will revolutionize the training of welders. These technologies will allow operators to practice GMAW techniques in virtual environments, simulating real-world welding scenarios without the need for actual materials or machinery [15]. This will not only reduce training costs but also improve operator skill levels, making the workforce more competent in handling complex welding tasks.

8.6.5 3D Printing and Hybrid Welding

Additive manufacturing (3D printing) combined with GMAW techniques could create hybrid processes that allow for on-demand fabrication and repair of metal components. Hybrid welding systems, which combine GMAW with other welding technologies, will be able to switch between processes in real time to adapt to different materials or joint configurations, offering greater flexibility and efficiency.

8.6.6 Sustainability and Eco-friendly Technologies

As sustainability becomes a key concern, GMAW technology will adapt to reduce energy consumption and material waste. Innovations in energy-efficient power sources, as well as the use of alternative gases and greener consumables, will minimize the environmental impact of GMAW processes, making it more eco-friendly without compromising quality or performance.

8.6.7 Conclusion

The future of GMAW lies in its integration with advanced AI, robotics, sensors, and digital tools. These innovations will enhance productivity, quality, and flexibility, making GMAW an even more indispensable tool for modern manufacturing and construction industries. As the technology evolves, GMAW will become smarter, more efficient, and more sustainable, ensuring its continued relevance in the welding world.

REFERENCES

1. R. Ruth, "Gas metal arc welding," in Welding Basics. Chanhassen, MN: Creative Publishing International, 2004, p. 44.
2. J. Norrish, "Evolution of advanced process control in GMAW: Innovations, implications, and application," Weld. J., vol. 103, no. 6, pp. 161–175, 2024. https://doi.org/10.29391/2024.103.015.
3. Y. Zhang, X. Wu, J. Gao, J. Chen, and X. Xv, "Simulation and ergonomic evaluation of welders' standing posture using Jack Software," Int. J. Environ. Res. Public Health, vol. 16, no. 22, p. 4354, 2019. https://doi.org/10.3390/ijerph16224354.

4. J. Eguia, L. Uriarte, and A. Lamikiz, "Analysis, optimization and accuracy assessment of special-purpose portable machines by virtual techniques," Int. J. Mach. Tools Manuf., vol. 111, pp. 31–42, 2016. https://doi.org/10.1016/j.ijmachtools.2016.09.006.
5. S. I. Wahidi, S. Oterkus, and E. Oterkus, "Robotic welding techniques in marine structures and production processes: A systematic literature review," Mar. Struct., vol. 95, p. 103608, 2024. https://doi.org/10.1016/j.marstruc.2024.103608.
6. N. Naden and T. J. Prater, *A Review of Welding in Space and Related Technologies*. Huntsville: NASA, 2020, April.
7. A. Greig, S. Rivas, S. Blackman, and W. Tizani, "Welding automation in space-frame bridge construction," Comput.-Aided Civ. Infrastruct. Eng., vol. 16, no. 3, pp. 188–199, 2001. https://doi.org/10.1111/0885-9507.00225.
8. K. Nogi and Y. Aoki, "Behavior of bubbles in welding for repairs in space," Mater. Des., vol. 18, no. 4–6, pp. 275–278, 1997. https://doi.org/10.1016/s0261-3069(97)00063-0.
9. Y. Huang, Y. Yuan, L. Yang, D. Wu, and S. Chen, "Real-time monitoring and control of porosity defects during arc welding of aluminum alloys," J. Mater. Process. Technol., vol. 286, p. 116832, 2020. https://doi.org/10.1016/j.jmatprotec.2020.116832.
10. M. Cerit, O. Kokumer, and K. Genel, "Stress concentration effects of undercut defect and reinforcement metal in butt welded joint," Eng. Fail. Anal., vol. 17, no. 2, pp. 571–578, 2010. https://doi.org/10.1016/j.engfailanal.2009.10.010.
11. I.-S. Kim, J.-S. Son, and P. K. D. V. Yarlagadda, "A study on the quality improvement of robotic GMA welding process," Robot. Comput.-Integr. Manuf., vol. 19, no. 6, pp. 567–572, 2003. https://doi.org/10.1016/s0736-5845(03)00066-8.
12. S. Penttilä, P. Kah, J. Ratava, et al., "Artificial neural network controlled GMAW system: Penetration and quality assurance in a multi-pass butt weld application," Int. J. Adv. Manuf. Technol., vol. 105, pp. 3369–3385, 2019. https://doi.org/10.1007/s00170-019-04424-4.
13. R. Thompson Martínez, G. Alvarez Bestard, A. Martins Almeida Silva, and S. C. Absi Alfaro, "Analysis of GMAW process with deep learning and machine learning techniques," J. Manuf. Process., vol. 62, pp. 695–703, 2021. https://doi.org/10.1016/j.jmapro.2020.12.052.
14. M. Tannous, M. Miraglia, F. Inglese, L. Giorgini, F. Ricciardi, R. Pelliccia, M. Milazzo, and C. Stefanini, "Haptic-based touch detection for collaborative robots in welding applications," *Robot. Comput.-Integr. Manuf.*, vol. 64, p. 101952, 2020. https://doi.org/10.1016/j.rcim.2020.101952.
15. V. S. Chan, H. N. Haron, M. I. Isham, and F. B. Mohamed, "VR and AR virtual welding for psychomotor skills: A systematic review," Multimed. Tools Appl., vol. 81, no. 9, pp. 12459–12493, 2022. https://doi.org/10.1007/s11042-022-12293-5.
16. S. Pattanayak and S. K. Sahoo, "Gas metal arc welding based additive manufacturing—A review," CIRP J. Manuf. Sci. Technol., vol. 33, pp. 398–442, 2021. https://doi.org/10.1016/j.cirpj.2021.04.010.

Index

For Product Safety Concerns and Information please contact our
EU representative GPSR@taylorandfrancis.com Taylor & Francis
Verlag GmbH, Kaufingerstraße 24, 80331 München, Germany